Yours is the Earth

Charles Mitchell, 1900

Yours is the Earth

THE LIFE AND TIMES
of
CHARLES MITCHELL

J. P. GABBEDY

PUBLISHED FOR THE ROYAL WESTERN AUSTRALIAN HISTORICAL SOCIETY BY THE UNIVERSITY OF WESTERN AUSTRALIA PRESS

First published in 1972
for the Royal Western Australian Historical Society
by the University of Western Australia Press
Nedlands, Western Australia

Eastern states of Australia and New Zealand: Melbourne University Press, Carlton South, Victoria 3053

Great Britain and Europe: International Scholarly Book Services Inc., London

U.S.A. and Canada: International Scholarly Book Services Inc., Portland, Oregon 97208

Library of Congress catalog card number 76-180214
SBN 85564 052 9

Printed in Australia by Frank Daniels Pty Ltd, Perth, Western Australia and bound by Stanley Owen and Sons Pty Ltd, Alexandria, New South Wales. Blocks made by Art Photo Engravers Pty Ltd, Perth, Western Australia

Registered in Australia for transmission by post as a book

This book is dedicated
to the unsung pioneers of the outback
and to the
Royal Western Australian Historical Society
in its efforts to orchestrate them

Contents

Illustrations

Errata

page 23, line 5, for 16 June 1870, *read* 16 June 1869

page 30, line 4, for 6 September 1870 *read* 16 June 1869

Preface

This story of my old boss and friend Charlie Mitchell, was conceived almost by accident. It was nearly six years in gestation and finally friendship was the main instrument of accouchement.

That it was ever finished at all, was, I think, due to my growing conviction, not only that Mitchell was interesting as a man but also that his life and work spanned a most fascinating period of the state's history which does not seem to have been recorded very fully.

On 12 August 1965 I succumbed to the blandishments of Dr B. C. Cohen, President of the Royal Western Australian Historical Society, at the luncheon tendered every year by the Commissioners of the Rural and Industries Bank in Perth, to commemorate the city's birthday. Dr Cohen suggested, and I agreed, that the historical records of the state must be infinitely poorer for not having recourse to biographies of people whose lives and work, although of great significance in our progress, have been so buried in the backblocks as to evade the scope of study.

One such person came to my mind immediately—Charles Douglas Mitchell. I had worked with him and had become aware, as time went by, of the profound influence his character had had on my philosophies. In the euphoria of the moment, I agreed to write 5,000 words about my old boss.

Ignorance is bliss—and just as well. If I had had any intimation of what I was undertaking I most certainly would never have ventured into the thorny thickets of biography. Nevertheless, it has been one of the most rewarding exercises of my life. The uncovering of hitherto unpublished diaries and letters led me on and on. Facts, figures and segments of unrecorded history were there for the finding. The weft of Mitchell's formative years—the wonderful home life no less than the strong outside influences—was inseparable from the woof of the times he lived in.

It was gradually conveyed to me that what might seem commonplace to many was actually the stuff of history, and I learned the sad fact that the writing and preservation of letters and diaries went out of fashion with the turn of the century.

In all, I interviewed 141 different people, some of them time and time again—their patience was inexhaustible. My one regret is that it is neither possible nor politic to print all the stories I have been told.

The result of those years of work and enjoyment I offer primarily as a tribute to Charles Mitchell, who was all in all a man. I hope also, however, that the book might be of some use to students, and that perhaps some line of my own inquiry might lead someone else to shed further light on a fascinating period of Western Australian history.

South Perth J. P. GABBEDY

Acknowledgements

No book of this nature could ever be written without the collaboration of many people, and it is perhaps one of the most pleasant tasks of all to record my deep appreciation of their assistance.

I am therefore forever indebted to Mrs Nora Rae, the last surviving member of Charles Mitchell's brothers and sisters—her old world charm and kindness made it an increasing pleasure, time and time again, to probe her remarkable memory.

I pay willing tribute also to John Honniball for his courtesy in reading the preliminary draft and then, so discreetly, explaining where research was needed and how to go about it. My special thanks go, also, to T. A. G. (Tom) Hungerford for his suggestions as to the format, his persuasion to write in the first person wherever it was apposite, and his insistence that I include a chapter on the Depression; under his guidance an amateurish essay became (I hope!) a readable biography.

Many other people helped—other than the 141 I interviewed—and it is simply not possible to pay to every one of them the tribute they warrant. However, among them, particularly, I am very grateful to Miss E. M. White, Librarian at the Department of Agriculture; to Miss Mollie Lukis and her staff at the Battye Library; to the *Narrogin Observer;* to Emmet Brown; Stuart Garland; M. E. Brinkworth; J. B. Mitchell and family; Major G. T. Mellor; George Grainger; Frank Johns; W. H. Cover; R. G. Courtenay; F. I. Barnett; Peter Doig; C. W. Edwick; C. G. Russell; Mr and Mrs Lu Cullen; A. A. Hardie; Sam and Stan Bradford; F. M. Collett; Clarrie and Kevin O'Connor; Mr and Mrs Hugh Henderson; W. Laskey of the Registrar General's Department; Eric de Luca and L. A. Jones of the Lands and Surveys Department; Miss L. Gardiner of the Nomenclature Committee; W. Y. Gannon, Principal Registrar, Mines Department; Mrs F. Ryan, Librarian; A. McGhie, O.I.C. Securities; and R. J. McKenna, Senior Economist of the Rural and Industries Bank;

erstwhile Bank staff-member, W. R. (Bill) Nelson, and my sequential secretaries, Misses Cheryl Cook and Sandra Miocevitch.

I was helped beyond the call of duty by the members of Charles Mitchell's own family—Joan, Nancy, Peggy and Bricky—perhaps of all people, I owe most to the lyrical Nancy, of prodigious memory, from whose notes so much inspiration and colour were drawn.

It is regretted that some of those mentioned above have not lived long enough to see this interpretation of their memories come to print.

Last, but certainly not least, I must thank my own wife Mary who, during the years of my writing this book, has had to put up with its deleterious effect on a temperament not noted normally even for an excess of sweetness and light.

1

Prologue

Bunbury, on the south coast of Western Australia, was named by Governor Stirling on 21 December 1836.[1] It is the state's second largest port and its largest country town, with 16,060 people;[2] it is the outlet for much of the rich dairying, agricultural, timber- and fruit-growing district of the South-West.

One wonders if its inhabitants realize, as the wool and wheat flood in from the hinterland, just how much of the present prosperity of the town, and that of the whole state, stems from the lifetime work of Charles Mitchell, who was born there ninety-two years ago, only half a century after the first slow ships intruded into the estuarine reaches of the Swan to found the Swan River settlement, and thirty-eight years after the first settlers moved into the Bunbury district buoyed up with the never-to-be-realized dream of Australind.[3]

In 1879 Bunbury was a very scattered settlement, extending from Haywood's Store at the shore end of the jetty to George Elliott's home at Bury Hill, now the site of the Catholic cathedral. It had a population of about 500 people: the 1881 census mentions 305 males and 278 females. James Moore, auctioneer and bailiff, was president of the Bunbury Town Trust. There were two medicos—Dr T. H. Lovegrove, colonial surgeon of Bunbury 1867-95, and John Sampson, surgeon. The cohorts of law and order—a police inspector, three constables and two water police under Magistrate W. P. R. Clifton—extended their protection to the settlements of Picton Junction and Australind. Spiritual needs were provided for by the Rev. J. Withers, the Anglican minister, and by Father H. Brady, the Catholic priest. Spiritous requirements—and in this it is accepted that supply always equals demand —were met by publican Richard Delaport and licensed victualler C. Wisbey.

Into this busy, self-contained and still isolated little community, on the quiet Sunday evening of 20 July 1879, a redheaded boy was born to William Bedford Mitchell and Caroline Mitchell—their seventh surviving child and fourth son.

The mother was by this time well experienced. Early that morning she had sent her eldest son James with a message for the midwife, Mrs George Lawrence, who lived with her carpenter husband in Princep Street, close by the Mitchell home, 'The Willows', in Stirling Street. Having attended the last nine births in the Mitchell menage, this buxom, competent lady was regarded almost as one of the family.

She arrived within the hour, promptly put Caroline to bed and sent a message informing Dr Lovegrove that he would be needed before the day was out. Her duties were not exacting, for the Mitchells at that time afforded a cook-housekeeper—a luxury dispensed with as soon as the eldest daughter, Kate, left school and began the normal training for a young woman in those days in keeping house. That time was yet to come, however, so after seeing the children fed and washed and sent to play, there was little for Mrs Lawrence to do except sit and talk to her patient and old friend—for whom she did the sewing, and whom she always helped to repaper the walls and ceilings.

It must have been a long day of gossip, for dusk had fallen when Mrs Lawrence despatched William, the husband, to fetch the doctor from his remote home at 'The Strand', at the far end of the town.

He went in the sulky, fitted with candle lamps. The roads were too muddy for safe walking and the only illumination was by kerosene lamps on each corner of Victoria Street. These were lit at dusk and dowsed at midnight, by which time most Bunbury folk would long have been in bed.

The children—James, Kate, William, Harry, Mabel and Amy—had been permitted to wait up, and when their new baby brother had been washed and dressed and laid snug in his cradle, they tiptoed in to see him. Harry and Mabel, the two redheads, were delighted to welcome reinforcements. They had been sorely teased by their dark-haired brothers and sisters, but now the odds were evening up.

The new baby was christened on the first Sunday afternoon in September when, in perfect spring weather, the whole family—father, mother and seven children—set out from the Willows for nearby St Paul's Church of England. Amy, aged two and a half years, was allowed to toddle the whole way and was delighted to do so, but eldest brother James, aged thirteen, was restless—he'd seen plenty of christenings, and thought them a waste of time on such a lovely day. He was repressed by Father's stern look and sister Kate's small hand on his coat-tail.

The godmother was Caroline's close friend, Mrs George Uglow, born Marion Earnshaw at Busselton in 1846. In her arms the baby was christened 'Charles Douglas' by the Reverend Joseph Withers. All lace and tucks in his six-foot-long family christening gown, he rested contentedly, nothing visible but the small puckered face and even smaller clenched hands which, one day, would play such an important part in shaping the destiny of the huge state of which he now was to become a registered member. That registration was the responsibility of the male parent and, somehow or other, William Bedford recorded his son's name as Douglas Charles. The error was not discovered until Charles came to obtain a copy of his birth certificate, as required for employment with the Agricultural Bank. Having then lived for nearly thirty years as Charles Douglas, and having served his country and married as such, he let it be, and, in fact, was to let it be for the duration of the long and crowded life that was to follow.

1. Previously known as Port Leschenault, after the botanist who accompanied the French expedition (known as the Baudin expedition) which examined the shores of Western Australia between 1801-04.
   Lieutenant Bunbury was a grand-nephew of Charles James Fox, the famous British statesman. He was attached to the 31st Fusiliers, and while stationed with a detachment of troops at Pinjarra, travelled overland to Vasse along the line of the present coast road, in August 1836. He was camped on the Bussells' property when Governor Stirling made a surprise visit in December. The Governor was so pleased with the young officer's report that Port Leschenault, of which Stirling had a high opinion as a port, was renamed in Bunbury's honour. Leschenault's name is perpetuated in our beautiful wildflower, 'leschenaultia'.
2. Census conducted in 1967.
3. The Western Australian Company, a large-scale immigration scheme, was launched in 1839-40 to implement Wakefieldian principles by settling the Australind district a few miles from Bunbury. The settlement was given the name 'Australind' in great hopes of developing a trade with India. The syndicate chose a site on the shores of Leschenault Estuary on Koombana Bay. The district was unoccupied and had originally been granted to Colonel Lautour, an absentee investor. The terms of the take-over of Lautour's interests are not known and the first record of a title is Crown Grant Enrolment

No. 447, dated 3 December 1844, naming William Hutt, Edmund William Jerningham, Henry Buckle, John Chapman, Charles Enderby and James Irving as Trustees for the Western Australian Company. The land, totalling 103,000 acres, was named Leschenault Location I, and later was altered to Wellington Location I.

2

The Antecedents

The birth of the redheaded fourth son of William Bedford Mitchell in Bunbury was, of course, subject to the same assiduous spinning as is devoted to all human affairs by the grey sisters in their cave, and it seems apposite now to examine briefly the events which led to the presence of Dr Lovegrove at 'The Willows' on that Sunday morning long ago, and beyond that again, what they decreed for the baby boy he then delivered: a shining life-long example of citizenship and thirty-five years of leadership in the evolution of Western Australian agro-economics.

William Bedford Mitchell, Charles's father, was born on 13 September 1833 at Penistone, in Yorkshire, of good yeoman stock. In the marriage register, nearly thirty-two years later, William was to describe his father as 'James Mitchell (deceased) Yeoman'. When twenty-two years of age, William sailed to Australia, in search of his fortune, on the 119-ton sailing ship, *Royal Charter*. He arrived at Melbourne on 17 April 1856 and proceeded straight to the gold diggings at Ballarat, but like many others, was too late. The gold rush had begun five years earlier and all likely sites had been well prospected.

Returning to Geelong, he obtained employment with a land agent, and in his spare time acted as the representative of his elder brother James who, in partnership with a Mr Namiby, conducted an export-import business back home in England.

We hear of him next from notes that he kept of a letter, dated 18 December 1859, written to his brother and mailed on the barque *Malta*, enjoining James 'that he was not to lose sight of the Swan River Berth on any account except he was doubtful whether I should fill it satisfactorily and if he thought so, in opposition to what I had previously said, and induced him to believe he could let the matter drop—as I did not wish him to recommend me further for the post if he had the slightest objection to do so, or if he himself was not convinced from what I had advised on the subject that he would be right in advising Mr. Prinsep to appoint me'[1]—an opinion which time was to prove over-pessimistic, and an attitude inherent in his yet unborn fourth son.

The 'Swan River Berth' referred to managership of the Prinsep estates situated in the Bunbury and Dardanup districts. The owner of the estates was Charles Robert Prinsep, Advocate-General for the East India Company at Calcutta.

Following a paralytic stroke in 1855, not long after the death of his wife, Prinsep retired to live in England, and Wallace Samuel Walsh Bickley of Fremantle became agent for the Prinsep interests in Western Australia.[2] In December 1853 Bickley's daughter Ida had married William Owen Mitchell[3]—no relation to W. B. Mitchell—the eldest son of the Reverend William Mitchell, an Anglican missionary sent to the Swan River settlement by the newly formed Western Australian Missionary Society, which provided the stipend.

On 18 March 1854 Bickley arranged for the appointment of his son-in-law, William Owen Mitchell, as resident manager of the Prinsep estates, in succession to Thomas Little, who had acquired land in his own right in the same neighbourhood. It was this W. O. Mitchell whom William Bedford Mitchell was to succeed in the position.

The story of how Prinsep became established in Western Australia is one which opens an interesting page in the history of the colony. There are two versions, which are undoubtedly related.

One is that William Locke Brockman of Herne Hill, sent to India by the Swan River settlers to purchase some cows to help stock up the colony, met Prinsep in the course of business. Brockman had purchased the cattle but was unable to arrange shipment, space being difficult to book and shipping being most infrequent. Prinsep then bought the cattle with a view to sending them to Tasmania, where he already owned property, but instead shipped them to Western Australia on his own account.

The other is that, seeking to spread his investments, he commissioned his Tasmanian overseer, Thomas Little, to go to Western Australia in 1838, taking cattle and horses, and there to acquire land on his behalf.

In 1839 Little decided to settle in the Bunbury area, possibly having met or been influenced by Brockman and further persuaded by Waller Clifton.[4] After landing, he purchased 1,832 acres on the tongue of land lying between Leschenault Inlet and the ocean—an area mostly of sandhills, with some rich flats along the shores of the inlet. On one of these flats, near the promontory known as Waterloo Heads, he set up a homestead, naming it 'Belvidere' after the Prinsep home in Calcutta. This was the base of operations whilst waiting for better land to become available.

Some time later Little purchased from W. Richardson of Bunbury a square mile of country (Wellington Lot 28) lying at the foot of the Darling Ranges to the east of Bunbury and traversed by the Paradise River. Subsequently he acquired also an area of 805 acres (Wellington Lot 61) south of Paradise, through which runs the Ferguson River. This he named 'Prinsep Park'.

In 1850 he purchased from James Henty, on behalf of his principal, 20,000 acres—being that part of the original grant of 60,000 acres for which Henty had received a title in fee simple—conditional on his relinquishing claim to the balance of 40,000 acres. This area adjoined Lots 28 and 61 to the east and was watered by Henty Brook and Collie River.

It was of these estates, assembled in the manner described above, that William Bedford was to become manager.

ON 17 FEBRUARY 1860 WILLIAM BEDFORD wrote to his family in England that they might expect him in the *Great Britain,* though 'I have not yet taken my passage which I suppose wouldn't be more than £20-25'. He did take it. S.S. *Great Britain* of 1,733 tons, master John Gray, sailed from Melbourne on 6 March 1860 bound for Liverpool, with 504 passengers, and in the passenger list Mitchell is described as: '28 years, single, miner, English'.

Although through brother James's representations managership of the Prinsep estates appeared assured, arrangements were not yet settled to William's satisfaction. In his letter of 17 February he went on to say that he scarcely understood James's meaning about the terms '£200 a year and 15 per cent on the produce yearly with £50 a year extra—providing Mr P. got good management'—not knowing whether James meant 15 per cent on the net or gross yearly produce of the run, nor whether he included the increase in live stock when he spoke of 'produce yearly'.

William arrived back in England in due course and on 27 July 1860 he wrote to Prinsep from 1 Belle Vue, Chelsea:

> . . . I shall be prepared and willing to go on the following terms: In order that I may be in a position to comply with any instructions you

may find it necessary to give concerning the disposal of the stock and produce, or general management of the property, during the above period, I should require an authority which would enable me to act without the interference, or being under the control of, any person employed by you in the care of the property, and also to order the payment, as you may direct, for the current disbursements of the estate. As regards salary, I consider that £200 would be a reasonable remuneration for the year . . . If after a few months residence on the property I should conclude and satisfy you that the estate could be made productive of profit whilst in your hands, I would then, as soon as circumstances would permit, inform you of the terms etc. on which I should be willing to remain for a further period of five or seven years, the remuneration then to be part salary and part commission . . .

Charles Prinsep wrote to his agent Bickley on 10 August confirming Mitchell's appointment, and a copy of this letter was evidently given to Mitchell as an acknowledgement to his letter. A further letter to Bickley, dated 28 September, *inter alia* gave further details of Mitchell's appointment:

Mr. William Bedford Mitchell whom I have selected to be my superintendent and to manage my Estate to the South, has taken his passage in the 'Prince of Wales' via Melbourne and will sail in a few days . . . This arrangement is to last for one year after which I shall determine what to settle permanently.

William sailed from Plymouth in the *Prince of Wales* on 16 October 1860, arriving at Melbourne mid-January 1861. The ship did not call at either Fremantle or King George Sound, as frequently was the case at the time. Possibly it suited William to journey this seemingly unnecessary distance; there may have been some unfinished business requiring his attention in Melbourne.

In any case, he wrote to Bickley from Melbourne soon after his arrival, advising that he proposed leaving by mail steamer for King George Sound (Albany) on 26 January. He would arrange passage to Bunbury, and as it was not William's intention to travel to Perth until he had completed a full inspection of the Prinsep properties and made his own assessment of the position, Bickley was requested to meet him on the Prinsep estates.

Bickley did not come to Bunbury to meet Mitchell. It is possible that the letter had not reached him, which was not an infrequent happening in those days. It is possible also that from the very vagueness of the arrival date, Bickley decided to take no action. A journey to and from Bunbury was a matter of consequence and would occupy four to five days. It was just as well he saved himself the journey, as William was two weeks overdue.

Mitchell recorded in his diary a letter he wrote to C. R. Prinsep on 18 February 1861:

> I arrived here on the evening of the 16th instant in an American whaler [*Elisha Dunbar*] and soon afterwards met up with Mr. McGuire, one of your tenants [believed the lessee of 'Prinsep Park']. I rode over with him to Paradise, 14 miles distant from Bunbury, the following morning and had a look around that place—which I found in a very rough state and apparently much neglected. I saw Slipper, the groom, who appears to be settling down to Colonial life a little better than he at first promised to do . . . is chiefly employed looking after the entires 'Commander' and 'Aakbar', and in handling young stock.[5] In the evening I rode over to Belvidere 21 miles from Paradise. Found this place in a rather better state of management than the other, but far from what it should be—said that this would have been a very valuable property now had it been properly cared for.
>
> Told Mr. P. that Hynes [overseer at Belvidere] informed me that he considered there was a sufficiency of wheat, barley and potatoes etc. in store to last until the next crops could be gathered. I also said that the American whaler had been detained by adverse winds at King George's Sound from the 1st to the 12th February and having rough weather after leaving port she did not reach Bunbury till the 16th inst. Asked Mr. P. to address my letters in future to Post Office, Australind, as it is several miles nearer here than Bunbury.

As the new superintendent of the Prinsep estates, Mitchell was having more than his fair share of management worries. While apparently he had been well briefed by C. R. Prinsep before leaving England, it now seems that those concerned in Western Australia were not fully informed as intended. This is indicated in two letters written in May 1861 which suggest that personal contact between Mitchell and Bickley had not yet been effected.

On 5 May W. B. Mitchell wrote to his predecessor, W. O. Mitchell:

> the appointment you seek to regain as local superintendent of Mr. Prinsep's properties is virtually [one] that I now hold myself and as I shall rarely be absent from the Estates—it is not probable that any other superintendent will be required during my stay here. You seem to write under the impression that I have taken the places on my own hands, but that is not the case and the difference, I presume, in the form of appointment between that which you formerly held and mine, is, that you received your directions from Mr. Bickley, while I hold my authority and get my directions from Mr. Prinsep.

On the following day he wrote to the agent, Bickley:

> You would learn from my note sent per Hynes that I wrote you on the 29th ult. in reply to yours of the 25th ult., but as the letter was

> posted too late, you will not receive it till you get this. First of all let me assure you that it is far from my wish that any misunderstanding should arise between us, in any way, with regard to Mr. Prinsep's properties or in any matter connected with them—and although you may think my letters are written in an uncouth style sometimes, I trust they will be none the less satisfactory on that account. If Mr. Prinsep did not acquaint you as to the authority he may give to me, or how I should stand in relation to yourself, I was led to understand he would do so, before I left England, in his letter to you dated 28th September last, and should the consequence of this result in his disapproval he will only have himself to blame for it.

No record remains as to how the situation was resolved but resolved it was, with W. B. Mitchell remaining in control of the situation and in an employment he would pursue for the next eight years.

EMBOLDENED BY A SENSE OF SECURITY, William began to look around for a wife, but it was not until 1864 that he lost his heart to Caroline Morgan, a very beautiful woman and an accomplished horsewoman whom he met for the first time at a picnic race meeting in Bunbury, while she was staying with her brother-in-law, Joseph Gardiner Reynolds, at 'The Island', Wonnerup.[6]

There was no lack of suitors for Caroline's hand, but William was persistent. He neglected no opportunity to visit the Island, and Caroline invariably rode through the lovely tuart and peppermint forest to meet him at Capel.

William was about thirty-two and Caroline just eighteen and a half when, on 24 July 1865, they were married in St Mary's Church, Vasse,[7] by the Reverend Henry W. Brown, colonial chaplain of the district.[8] The witnesses were J. G. Reynolds (brother-in-law), C. L. Morgan (brother) and Hannah Hough (sister-in-law). On the distaff side, the blood lines of the redheaded fourth son who would be born to them almost fourteen years after the memorable day in the little church in Vasse, reached back to solid Welsh working stock, strengthened with the proud independence of English yeomen.

Caroline's father, John C. Morgan, was born in Wales in 1817. Sponsored for passage in *Parkfield* by M. W. Clifton under the terms of the Australind Land Settlement Scheme, he is described in the passenger list as 'single man . . . steerage passenger . . . carpenter and joiner'.

A shipboard romance developed with Maria Laura Feast, 'widow . . . servant . . . aged twenty-five years', and the couple were married at Bunbury on 29 August 1841 by the Reverend J. R. Wollaston, a Church of England minister who had come to Australind in May of that year. They had four children: Charles Llewellyn, born 5 July 1842 (married Sarah Hough) ; Mary Ellen, born 26 July 1844 (Mrs Joseph Gardiner Reynolds) ; Caroline, born 19 January 1847; and Laura Maria, born 10 October 1850. The three elder children were born at Australind, Laura Maria in Bunbury.

Mrs Nora Rae, the fourth daughter and the last survivor of the Mitchell children born at the Willows, recalls that Caroline, her mother, used to point out the old Australind church when driving past in a sulky to visit the Cliftons, saying that she and Aunt Mary Ellen had been born there, but neither adding nor implying they had lived there.[9]

Historical sources are not clear as to the actual purpose of this building when first erected. Undoubtedly built by Clifton, it may have been some sort of storehouse, and while 'Upton House' was being built it might well have been used as a temporary dwelling, and quite likely for such special circumstances as a confinement and the subsequent lying in. Accommodation for the workmen and their families was in tents or bark huts.

Probably it was in this building that Reverend Wollaston conducted his Sunday afternoon services. Mrs Clifton, a masterful woman of Quaker persuasion, would have been most unlikely to have made the minister welcome in Upton House, or to have permitted the use of her own 'house of prayer'. On the other hand, the stern unyielding old man would never have permitted himself to be beholden to Mrs Clifton. Thinking poorly of any branch of religion but his own, he could have also entertained generally hard thoughts of the Cliftons, from whom he had hoped to receive the appointment as chaplain to the Western Australian Company, with accompanying stipend and support in the building of a church.[10]

Although it is suggested that Morgan had had some training in architecture and described himself as 'builder', the nature of his indenture to the Cliftons allowed him little scope for his talents. It is not known when the shackling obligations were discharged, or exactly when he struck out on his own, but it must have been some time after the completion in 1847 of the Clifton's new home, Upton House, on which he is known to have worked.[11]

The Western Australian Company having failed, building at nearby Bunbury came to a standstill. With only labourer's wages offering to support his growing family, Morgan became a licensed victualler.

In a history of early Bunbury, written in 1946, G. W. Clarke records:

> It would appear the Hotel business was not profitable in 1850 and a number of licenses were allowed to lapse. In that year, George Elliott, the Govt. Resident, in reporting to the Governor, stated: 'I have to inform you the only Publican's license was granted to Mr. John Morgan of the Bunbury Hotel.' It will be noted that the word 'Hotel' is used, such premises being usually known as Inns.
>
> As to whether Morgan took over one of the existing licenses or not we cannot state, but older residents maintain that Morgan conducted a Hotel in Clifton Street and this was probably the Bunbury Inn referred to.[12]

The hotel for which Morgan held the licence from 1849 to 1852, and again in 1856, was apparently no more than a beer shop, with no provision for either meals or accommodation. On his second archidiaconal visit to Bunbury (20 March 1851) the drink-hating Archdeacon Wollaston wrote: 'Wretched Bunbury, no more houses but more drinking than ever. Ramsey with grog shop at Child's[13] in opposition to Morgan[14] at the Hotel.'[15]

Morgan, however, was sensitive enough of the shortcomings of his premises to lodge his family some distance away. At the conclusion of his first period as licensee of the Bunbury Hotel, perhaps prompted by the stagnation in the town and perhaps to take advantage of a growing timber trade and a free licence,[16] he leased from James Lawrence a piece of land at Picton, just above the only ford available to timber hauliers crossing the Preston River to Dardanup (today the site of a modern bridge), through the hills to Ferguson and on to Collie. There, mainly by his own efforts, he built two large huts from handhewn jarrah slabs, the chinks filled and inner walls plastered with mud mortar containing a high proportion of clay with burnt limestone, the mixture often referred to as 'pompey'.

One hut was used as an inn, but reports differ as to whether Morgan's family resided in the second.[17] However, because of the distance from Bunbury, difficulty of transport, and necessity for constant attention to his business, it may be accepted that the family accompanied him to Picton.

The adult Caroline's complete aversion to alcohol in all forms, and her persistent campaigning for its abolition, did not arise from any inhibition about her father's trade, but from a later association. She always spoke of her father as 'a complete gentleman', considerate, courteous and protective.

The timber hauliers made the Morgan inn a night staging post; they could cross the ford only in daylight. The roistering scenes can be well imagined and it is a wonder that neither the inn nor its 'occasion of sin' was ever mentioned by Wollaston, who would have had to use the nearby ford during his 1853 and 1854 tours. Its existence must have been specially galling to him as the inn was built on ground he once owned within a half mile to the east of his beloved St Marks. Thankfully, neither building was visible to the other, nor were the drinkers' sodden matutinal slumbers disturbed by any sabbath bell—St Marks did not possess one. It could be assumed that, wisely, drinking would have been conducted at a discreet level whilst church services were in progress.

By the end of 1855 the timber industry was at a low level. Although markets were being sought, Morgan could not afford to await the upturn. His wife's health was deteriorating, and medical attention was available only at Bunbury. He returned to the hotel in Bunbury at some time before March 1856, for Wollaston recorded of those at a service held on Sunday 2 March in Bunbury's temporary church (used also as government school house) : 'Mrs Morgan, in last stages of decline, poor woman (hers was my first marriage in the Colony—visited her afterwards) '.[18]

Commenting as usual on one of his pet aversions, Wollaston continued: 'the drinking end of Bunbury (as I call it), Child's former store and Knight's house still kept by Ramsey and Morgan respectively—as miserable, dirty and rubbishing as ever'.[19]

However, he recorded his approbation of the 'fine new jail at the back [of Wollaston's residence] near the Barracks towards Bury Hill',[20] at the junction of Stephen and Wittenoom Streets.

Strange he should have mentioned it. At the time of its demolition in January 1969 it is recorded as having been built by John Morgan in 1848 and upgraded in 1852 following the introduction of convict labour.

Maria Laura Morgan died of double pneumonia on 23 November 1856, and her husband followed her on 9 August 1858. After Mrs Morgan's death, Caroline and her sisters were cared for in the home of the Reverend Henry Brown, a graduate of Oxford, who had arrived at Fremantle on 13 September 1852 in the sailing vessel *Samuel* of 327 tons. His first posting was to Bunbury, where a parsonage was being built, and shortly after his arrival he married Lucy Clifton. Unfortunately they were never blessed with children, and Caroline received in full their unstinted love and care. She remained with them until her sixteenth birthday, when she went to live with her sister, Mrs Reynolds, at the Island and so set in motion the events which led to the appearance of our redheaded hero at the Willows in Spencer Street, Bunbury, in 1879.

C. R. Prinsep died in 1864, a year before William Mitchell's marriage. A brother, Henry Thoby Prinsep, formerly military secretary to the Bengal government, was appointed his executor. In a letter of 21 December 1864 Mitchell acknowledged receipt of a power of attorney and copy of the will, which he had passed on to George Leake, a solicitor of Perth, to arrange local administration on his behalf.

The friction which seems to have developed under the executorship, and which had come to a head by the end of 1865, might have been caused by horse-racing.

The principal purpose of the Prinsep estate at that time was the breeding of remount-type horses for the Indian Army and the Near-East markets. Then, as now, one of the best methods to popularize a stud and to be able to command higher prices was to produce a champion race-horse. Mitchell had trained a gelding, Star of the South, reputed to have been bred from either a Chuson or Sir Robert colt, and which as a six-year-old, ridden by Stephen Parker,[21] won the Queen's Cup over 3 miles in 1865.

At this meeting, Star of the South won a double and Mitchell's Southern won two out of the three heats in which he ran. The Queen's Plate was the colony's premier racing event. It was first held in April 1853 with a stake of £50.[22]

Mitchell wrote rather tersely to the executor on 24 January 1866:

> Enclosed you have a statement showing the result of my agency. . . . With regard to Mr. Bickley having been more punctual than me in rendering regular and frequent accounts I admit . . . this superiority in corresponding with you and all that; but examine our respective accounts and judge . . . whether he has succeeded better than me in the actual management. He left the Estate liable to Mr. C. R. Prinsep

in the sum of £1,949. 7. 6—I, up to the end of December, show that its total liability after including my salary of £1,050, is £536. 9. 10, as the account will show and this is as near as I can estimate everything. This is without valuing increase of stock, cultivation and improvements—which according to Mr. Bickley's valuation would exceed £3,000. . . . I will merely say . . . that I have succeeded in dealing with the Estate better than any one of my predecessors . . .

As for betting or making matches in racing you need not imagine for a moment that I ever supposed you would have anything to do with anything of the sort nor interfere with me in it either; but for the information of yourself and your nephew before he comes out it may be as well to say that the extent of my business in this way was to win £5—in bets on Star of the South, and lose £10 in the matter of running another horse—so horse racing has not been very profitable to me. I maintain it is unfair treatment of me to claim the winnings of Star of the South and I am certain you would be more liberal if you were here to arrange the matter with me on the spot.

The longer your nephew delays coming here the more detrimental it will be to his interest I am afraid—for there is a large amount due to the Estate and I doubt if my present power to recover is not sufficient to some cases—such as distraint for rents etc. As my accounts are to be passed by your nephew I do not send you detailed accounts of receipts and disbursements, but perhaps it may be better to do so soon, as I am anxious to receive a settlement of my salary and the balance due to me on balance of Current account as I am not in a position to furnish funds wherewith to carry on the Estate. I also beg to give you notice of my intention to claim my passage back to England and on my arrival there I hope to have an opportunity of seeing you concerning matters relating to this Estate . . .

I have taken up 7,000 acres of land as a squatting run, for an annual License of £7. This land adjoins the Henty Grant and it is taken from the beginning of the year for twelve months as an addition to the Henty Grant and to prevent other parties taking it. A deal of land has been taken for sheep farming lately in this district and I was afraid this might be required for the purpose and if it were the sheep in time would drive other stock off their run . . . We have not had a drop of rain for nearly three months now and the feed is very much burnt up and water very scarce. The weather has been hot, but cooler than last season.

We are now busy thrashing and I fear the crops will be very light. This is the first season that cash has been readily procurable for produce since I have been here.[23]

Two months later, on 31 March, Mitchell again wrote to the executor.

It is well that Mr. Henry C. Prinsep will be here soon for many reasons and I hope he will not make a very long stay at Singapore. We want tenants for both the McGuire's farm and Belvidere, but I have not tried to let them on lease until his arrival.

> The agent at Singapore wrote me by the last mail saying that horses are in good demand—light animals such as they need here—but your nephew will learn all about that market and their requirements being right on the spot.[24]

The nephew who had been appointed manager for his father's trustees was Henry Charles Prinsep, born at Calcutta twenty-two years before. He attended Cheltenham School in England and was a man of most artistic taste, and a cousin of Val Prinsep, a well-known English painter. One of his works—a black and white etching of W. R. Mitchell and himself, duckshooting during a trip to Albany by horseback—is in the possession of Guy Mitchell, of 'Sylvan Loch', North Dinninup.

Henry arrived in Western Australia in May 1866 and immediately displayed more interest in the social life of the colony than in his managership. There was so much to see that excited him and one can well imagine how an eligible bachelor of his attainments was made welcome. Invitations poured in from all sides but it was Josephine Bussell of 'Cattle Chosen' who captured his heart. They were married a year after his arrival and on the eve of his wedding day he happily told his friends: 'Today I am in Busselton but tomorrow I shall be in "Paradise" with Josephine.' The marriage had confirmed his decision to remain in Western Australia and he immediately set about building a fit residence for his bride, a delightful brick and stone home of ten rooms which stands on Dardanup Lot 21, owned today by F. L. B. Craig, rightly preserving the name of Prinsep Park.

Henry, however, continued to take little interest in the farming and horse-breeding which had created his assets. He leased all his farms as soon as possible, and cattle were disposed of in favour of sheep, which proved quite unsuitable to the damp flats. He occupied himself as buyer and exporter in the trade of sleepers, for which the price—at landing—was 4s 2d each, subject to inspection.

In 1870 Henry lost by shipwreck a full cargo of sleepers and horses—unfortunately not insured—consigned to India. His financial position was further aggravated by a continuing drop in the price for horses and he was forced to relinquish the estates in 1874. They were purchased about 1877 by W. H. Venn, who had also acquired the Dardanup estate of Thomas Little. Little, like so many of his contemporaries, had fallen into debt beyond his assets. The estates were eventually subdivided under the direction of the Western Estates and Timber Coy, of London, and became Dardanup Estate Lots Nos 1-29.

One dwells upon the affairs of the Prinsep family not only because they are relevant to the life of William Mitchell—and so, by extension, to that of his son Charles—but also because they reflect some interesting aspects of Western Australia's development, and depict one aspect of the rise and fall of the landlordism and tenant farming, with its sociological and economic shortcomings, on which this state was founded.

Land grants, purporting to secure to the holder a comfortable and profitable future, far exceeded in size grants made in the early settlement of the eastern section of our island continent.

The policy was designed as an incentive to introduce much needed capital and personal on-the-spot involvement. There was then no other known short-term method. However, what it produced in effect was a colonial version of the English style of landed gentry, the self-appointed genteel members of which established a meticulous grading of social distinction, and a patronizing attitude to all outside their narrow circle.

These attitudes, not necessarily applicable to or applied by the Prinseps, were repugnant to William Bedford Mitchell, whose yeoman ancestry recognized caste only as a result of honoured service. To him freehold farming land was a trust, to be used fully and wisely. He himself had experienced the wide gap which divided the rights of employer and employee—it was not till 1900 that legislation placed master and servant on a footing of equality in the courts.

He also witnessed the workings of the convict system introduced into Western Australia in 1850, with shipments continuing until 1868. Hirings were permitted after the convict had served a probationary period, usually of six months.

The agreement to accept convicts into Western Australia was not only to provide an urgently required work force, but in return the home government undertook, at its own expense, to introduce an equal number of free settlers.

Within five years sufficient labour, free or forced, was available for the first time for the three great needs of the community, to cultivate the land, construct improved means of communication and transport, and provide a market for the produce.

After 1868 the number in prison, on public works or in private service declined slowly. Not until 1886 was the imperial convict establishment finally dissolved and the remaining convicts (200) placed under the control of local government. The introduction of convicts did nothing to alter the iniquities of the master-servant system. The coercion permitted in respect of the hired convict or ticket-of-leave man encouraged similar treatment of other servants.

The Western Australian authorities considered they had been let down in the overall arrangement. By counting two children as the equivalent of one adult, they argued that only 6,122 free settlers had been introduced against a total of 9,721 convicts. Nothing came of their protests and representations.

Too great a proportion of the granted land lay idle or was inefficiently farmed. In 1829 Edward Gibbon Wakefield, without either experience or first-hand knowledge, propounded principles of colonization that attracted wide interest at the time. His fundamental assumption was that the natural tendency among labourers, desiring independent status as landowners, was to work hard as tenant farmers, and save their money towards purchase from the grantee. This proved wrong in practice. There seems to have been a dearth of the two main spurs, example and practice, probably simply because the employing class could not supply them.

'It is unfortunate that large grants were made on basis of capital goods', Margaret McAleer observed in 'Some aspects of the development of the Northern Wheatbelt' in 1956. 'Too often the prospective settlers had disposed of most of their ready money, and too often for costly articles of civilisation rather than for necessary implements of stock which would have been of first consequence in the Swan. Such knowledge as they had of farming had therefore to be transposed very low indeed.'[25]

When Sir John Forrest's government came into power, with the grant of responsible government to Western Australia on 29 December 1890, it was patent that the system of settlement through land grants had failed. Sir John was well aware that the freehold tenure of large land grants generally had led to lazy utilization of them, which was certainly not in the national interest. He was equally aware that leasehold, especially if its term is short, leads to low-grade exploitive occupancy and inversely, that in farming economic efficiency is markedly influenced by the way in which the land is held, and the attitude of its owners.

Some years later, added emphasis was given to earlier legislation in the passing of the Agricultural Lands Purchase Act, assented to on 27 October 1896: 'An Act to make provision for the purchase of lands suitable for immediate settlement and for facilitating the better settlement of the people on the land.' The act provided for the making of a contract for the acquisition of land by surrender at the price fixed by the Land Purchase Board as the fair value thereof, or at any lesser price.

The break-up of large idle estates, particularly of those near a railway, had commenced. The old order had been replaced with the most liberal set of land laws in the British Empire. Virtually unaltered, they are still extant, designed to promote private ownership in freehold, on easy and liberal terms, giving equal opportunity to both rich and poor.

William Mitchell was deeply affected by the history and shortcomings of landlordism, by tenant farming, the social derogation of the servant, and the exploitation of convict labour. His recounting of it evidently had a strong impression on young Charles who, in later life, not only detested snobbery, but became dedicated to the principles of private ownership and the right of equal opportunity. He also strongly opposed all forms of exploitation—not only for the preservation of the rights of the individual but to their being preserved in spite of the individual's self.

William Bedford, steeling himself to meet and greet young Prinsep without any display of rancour, found to his delight that he liked the quiet natural new boss, unassuming in his proprietorship. The admiration and respect became mutual, and William stayed on as friend and adviser, without diminution of salary or loss of status.

The quarters at Paradise were comfortable, and there James was born on 27 April 1866, and Kate the following year on 15 April.

With the gradual leasing of the various Prinsep properties, William found his responsibilities lessening, and late in 1869 he suggested in all conscience to Prinsep that his services could well be done without. Henry could do little else but concur, much as he hated to lose a close and trusted friend, but he was reassured when William offered his services and advice at all times. William intended setting up as a stock agent, centred on Bunbury; he would pass through Dardanup often in the course of his work, and would always 'call in'.

1. From William Bedford Mitchell's unpublished diary, owned by Bremer Mitchell of 'Uda Bremer', Donnybrook, who has consented to the diary being lodged with the Battye Library on loan on publi-

cation of this book. All subsequent letters quoted in this chapter are from Mitchell's notes for, or copies of, letters in this diary.

2. Bickley arrived at Fremantle in the sailing vessel *Protector* of 380 tons on 25 February 1832, apparently to take up the position of Lloyd's agent. Later he became the representative of several foreign insurance companies, commercial houses and shipping firms. He was appointed a Justice of the Peace, and became a Legislative Councillor from 1872 to 1875.
3. William Owen Mitchell was the grandfather of Clayton Clealand Bickley Mitchell, elected M.L.A. for Stirling on 31 March 1962.
4. Marshall Waller Clifton, chief commissioner for the Western Australian Company, accompanied by his wife Elinor and ten of their family of fifteen, arrived with the first party of migrants at the Leschenault Estuary in the barque *Parkfield* of 600 tons on 18 March 1841 to settle at Australind (see n. 3, Chap. 1). The *Diadem* brought further settlers and following the arrival of the *Trusty* in December 1842 with the final group of migrants, the company ran into financial difficulties and the project was abandoned. Less than 500 people were brought out, but numbered among these were the forefathers of several men who were to play a prominent part in the later history of Western Australia, most notably Alexander Forrest, John Forrest and Newton Moore.

   Clifton was first placed on half-pay, and received notice of dismissal late in 1843.
5. Slipper had arrived from England in May 1860 in charge of a blood stallion and a well-bred mare.
6. Following the murder of John Hurford by the blacks in 1860, Reynolds disposed of his Tanjanerup and Dundinalup estates on the Blackwood and purchased from Hurford's beneficiaries 'The Island' property, situated between the mouths of the Vasse and Wonnerup Estuaries. It is now occupied by Alfred Reynolds, a grandson of J. G. Reynolds.
7. The address Vasse, as given in the Marriage Register to St Mary's Church, is intriguing. The name was first recorded by the French expedition of the ships *Geographe* and *Naturaliste* under the command of Nicholas Baudin, who visited these shores in 1801, in memory of a sailor named Vasse, who one night was lost in the heavy surf and presumed drowned. The townsite at the Vasse was given the name of Busselton in honour of its first settlers, the Bussell family, on 4 July 1835, and a general notice to that effect was issued from the Surveyor General's office in Perth. The term Vasse, however, continued to be used generally until the end of the century.
8. The term 'colonial chaplain' indicated appointment by the Governor, and the inclusion thereby in the Civil Establishment; salary was £200 per annum paid from colonial and convict funds.
9. Lands Titles Office records disclose that the land, portion of Wellington Location 48, now of 11 5/10 perches, the subject of certificate of Title Volume 621, Folio 96, was sold by the Western Austra-

lian Company to James Narroway some time prior to 1860. Narroway then donated the land to the Congregational Church for a chapel site. The application for a Possessary Title, lodged by the Augustine Congregational Church, Bunbury, incorporated in 1914, recalls the chapel was erected (aparently renovated or modified) by one John Gibbs, prior again to 1860. It is possible that the land, being part of an early grant, could have been disposed of quite legally in those days without record being maintained. On issue of fresh Title, C/T 621/96, the land and building, through the aegis of Bishop Frewer, was acquired in 1914 for £30 by the diocesan trustees of Bunbury Incorporated. The church was licensed on 22 December 1915 and renamed St Nicholas Church.

Memorials to Marshall Clifton and his wife are retained in a plaque erected on the site to their memory: 'To the Glory of God and in memory of MARSHALL WALLER CLIFTON FRS, Born 1787, died 1861, leader of the band of Pioneers who, in connection with the Western Australian Company landed at Australind March 18, 1841, and of ELINOR, his wife, Born 1792, died 1866, who bravely shared with her husband the trials and difficulties of those early days.'

10. Rev. John Ramsden Wollaston was born on 28 March 1791 in England. Inspired by Captain Stirling's advertising the merits of Western Australia as a new field for colonization, he accepted the challenge, gave up a comfortable living, and at the age of fifty he migrated with his wife, three of his five sons and two daughters, arriving at Fremantle in the *Henry* on 19 April 1841.

After some weeks at Fremantle he took passage in the cutter *Venus* for Port Leschenault, landing at Point Casuarina. To give shelter and sustenance to his family, he purchased 115 acres from the captain of the American whaler *Samuel Wright*. Five acres had been cleared for a garden and some huts of doubtful cleanliness erected.

On the Western Australian Company's declining to utilize and pay for his services, Wollaston's only alternative was to apply to the government for appointment as a colonial chaplain. The appointment was promised, but made conditional on the completion of a church at Picton. Wollaston's private means were soon exhausted, but by superhuman efforts and the suffering of extreme privation, the church was completed and opened on 18 September 1842. He was then appointed to full-time clerical duties at Bunbury where he remained until 1848. He purchased Leschenault Location 39 of 100 acres from the Crown on 16 April 1844 for £100. At the government's request he later agreed to take charge of Albany Parish and in 1849 was created an archdeacon. He died 3 May 1856.

11. Named after Upton Place, England, where Mrs Clifton once lived. The main building, in brick, mostly two storeyed, still stands and is in excellent preservation, thanks to recent skilful restoration. The present owner and occupier is R. Bingham Clifton, a great-grandson of M. W. Clifton.

12. G. W. Clarke, *Early History of Bunbury* (Bunbury, 1946).
13. J. K. Child built the first brewery in Bunbury, which at the time found favour with Wollaston to the extent that beer was less potent than rum.
14. Morgan's 'Hotel' was referred to as 'Knights'. James Knight apparently took the first licence for the building in 1846, in which year licences were also recorded for Hugh McDonald and Hugh McKenzie.
15. *Wollaston's Albany Journals, 1848-1856,* collected by Canon A. Burton, edited by Percy U. Henn (Perth, 1954), vol. 2, p. 117.
16. A free licence, apparently not subject to official record, could be granted an innkeeper who provided some type of public service, e.g. feeding or housing the mailman and his horse, and whose net income did not exceed a certain figure.
17. Built on Portion of Leschenault Location 39 Diagram 628, plan 2144 of 1 acre 3 roods and 3 perches. Location 39 was purchased from the Crown by the Reverend John Ramsden Wollaston on 16 April 1844 for £100. The whole of the land (100 acres) was sold to Harry Sillifant of Leschenault on 6 April 1850, who in turn sold that portion adjoining the river on the south-east corner to James Lawrence later in the same year.

    The actual inn still stands, although the 'residential hut' was removed to make room for a modern dwelling occupied by the manager of CSBP and Farmers Ltd, whose company purchased the land and buildings on the establishment of their works at Picton Junction. The inn was in full use, run mainly by the Lawrence family, until the Wimbridge family built their hotel in the townsite, opposite the railway station, in 1901.
18. *Wollaston's Albany Journals,* vol. 2, p. 117.
19. *Ibid.* p. 235.
20. *Ibid.* p. 234.
21. Stephen Henry Parker was born at York 1846. He was an amateur rider, who studied law, and was admitted to the Bar in 1867. Entering politics in 1878, he remained in close association till 1897. The first native-born Western Australian to be appointed Chief Justice, 1906, he was knighted in 1908, K.C.M.G., K.C.
22. It continued until 1879 when it was overshadowed by the Metropolitan Handicap which became the Perth Cup in 1887.
23. Extract from a letterbook of W. B. Mitchell, 20 June 1863–31 March 1866, the property of Mrs Nora Rae.
24. *Ibid.*
25. Margaret McAleer, 'Some aspects of the development of the Northern Wheatbelt', *Early Days, Journal and Proceedings of the Western Australian Historical Society,* vol. 5, part 2 (1956), pp. 45-70.

3

The Family at 'The Willows'

Late in 1869 William and Caroline moved their family to Bunbury and while their new home was being built, took up residence in Prinsep Street, in a cottage owned by the Newton Moore family. It was while they were living there that William Bedford, their second son, was born on 16 June 1870.

Mitchell had purchased one acre on the western side of Victoria Street between Stephen and Stirling Streets, subsequently subdivided into Lots 222, 223, 224 and 225. Here, on what was later Lot 225, he built 'The Willows', named after two large willows growing nearby.

A ground plan is recorded in his diary. A frontage of 40 feet 6 inches and a depth of 26 feet enclosed the main building of five rooms and a passage-way. There were 6-feet-wide front and side verandahs, and a back verandah of 5 feet. A wing 20 feet by 12 feet, containing two rooms, extended from the rear on the northern section. G. Floyd was the contracting carpenter; there is no record of who did the brickwork and plastering. The shingle-sheathed roof was so well done that it persisted until demolished in 1962. The cost of the work was recorded by W. B. Mitchell in his diary:

Brickwork 385 2/3 yds @ 1/4	£25. 14. 4
Plastering 4198 ft = 466 4/9 yds @ 6½d (2 coats on walls)	12. 12. 5
Plastering 515 ft = 57 2/9 yds @ 1/5 (ceiling — 3 coats)	4. 1. 0
Allowing foundation to be 14 inches wide and 2 feet deep, all this including for 4 fireplaces (40 ft) then will be 555 running feet = 61 2/3 yds @ 1/9	5. 7. 11
	£47. 15. 8

Twelve more children, eight of whom survived infancy, were born to William and his wife at the Willows between 1870 and 1888.[1] To accommodate the family, the northern wing was duplicated on the southern rear section. A Dutch oven, built of mud and clay, for the baking of bread or extra large joints of meat, occupied a prominent position in the back-yard. Stables, a shed for the house cow, and also a large fowl-yard were built on land that was to become Lots 223 and 224. A natural spring bubbled to the surface on the future Lot 223, providing year-round stock water.

Looking back over the intervening years Mrs Nora Rae recalls her childhood home as most comfortable. Its only fault was that, following the English style, the verandah practically abutted the footpath. However, being at the end of the main street, it was delightfully situated for visitors, and became a meeting place of the town. The Mitchell's hospitality was a by-word of the day and Kate's fine piano playing a relaxation to the guests.

The journey home from shopping and other expeditions in 'town', now merely a quick trip by car, was then a matter of many hours of often uncomfortable travel, by buggy or any other kind of horse-drawn vehicle which rattled and clattered around the embryonic roads of the South-West: for instance, the Cliftons might have taken an hour between Bunbury and their home. None of the Mitchells' many friends, particularly the Timperleys (Resident Magistrate) and the Waller Cliftons, left town without calling in for a cup of tea.

After the subsequent subdivision, the corner block, Lot 222, was sold to a Stephen Smith of Sydney in 1883, for £500. Subsequently this block was acquired by the Commonwealth Bank, which also in 1949 acquired by compulsion Lot 223 for £2,975.

The Commonwealth government required the remaining blocks. It had acquired Lot 224 for £5,280 in 1953, but was forced to wait until the death of Kate, at the age of ninety-three in 1960, before it could purchase Lot 225 for the vastly increased price of £17,000.

In this respect Kate, with whom her sister Amy lived until her demise, proved a considerable embarrassment to the Commonwealth government. She had become virtually the 'grande dame' of Bunbury. Imperious, of prodigious memory, and a stern lifelong arbiter of moral and social behaviour, Kate refused to bow to senility and remained an alert and lively person until eighteen months prior to her death, when she was confined to her bed.

She was one to be handled with all due caution and the Commonwealth government, for once, realized that it had met its match.

Amy, on the other hand, was a much quieter lady, who conducted a flower shop and worked untiringly for her Church and patriotic appeals.

IN PURCHASING CATTLE FOR RESALE on the metropolitan markets, William Bedford travelled widely, as far afield as the 'Deeside' Muirs, east of Manjimup. His constant and faithful companion was a native stockman, 'Tommy Coolbull', who, incidentally, survived his master by less than a year.

His horse-trading, while contributing in no small measure to his income, led him into several very unprofitable ventures. There was a strong demand in India for cavalry-type remounts, which Mitchell bought where he was able, and consigned by sailing ship to India. The horses, harnessed into a broad canvas surcingle with a metal ring fitted to the top, were swum out to the ship anchored in Koombana Bay. Once alongside, they were hand-winched aboard with a block-and-tackle fixed to the end of a lower yard-arm.

Storage of fresh water was limited and any undue delay on the voyage because of unfavourable conditions could be disastrous. Horses died or had to be destroyed. Three consignments of Mitchell's horses were caught in protracted equatorial doldrums, each occasion bringing him to the verge of bankruptcy. Twice he was saved only by the intervention of his son Harry with the loan of moneys and a personal guarantee to his father's bank for overdraft accommodation.

The other occasion concerned a purchase of remounts which were bred by Maitland Brown on his 'Glengarry' and 'Newmarracarra' properties in the Greenough Valley, 300 miles north of Perth, and loaded at Dongara. The Brown family had pioneered the India remount trade and by the 1860s enjoyed practically a monopoly. However, on being advised of the complete loss of the cargo, Maitland Brown generously deferred his claim, advising Mitchell 'to pay him when he could'.

In spite of these setbacks, William Bedford still prospered. He had observed and taken a great fancy to the heavy red soils around Donnybrook and foresaw a great agricultural future for the district. In the early 1890s he purchased the freehold of the Railway

Hotel, on the site the hotel now occupies, and held it till his death.

Mrs Rae holds the conviction that her father became wed to the district when he acquired an interest in Leschenault Location 55, the original holding taken up by Nash and a party of young Irishmen in 1842 and named 'Old Donnybrook', from which the town eventually derived its name. However, no official record of William Bedford's interest can be found.

There were two other hotels in Donnybrook—Trigwell's Anchor and Hope and Fred Lindrum's Donnybrook. It was in this latter hotel that Fred's sons, Fred and Walter, later to become the world's champion snooker player and champion billiard player respectively, continued their primary grounding, standing on packing cases to play their shots.

Initially, William Bedford's office was in the front room, or drawing room of the Willows, where his paper-littered table and a filing cupboard came to be accepted as part of the furnishings. With the Perth coach stopping overnight in Bunbury, clients and friends came and went at all hours. They were invariably made welcome: the Mitchells kept open house, and there was hardly a meal without a visitor.

William Bedford's stock dealings gradually tapered off following the turn of the century, partly because age prevented his travelling as widely as he had, and partly because of an increasingly competitive buyer's market.

In January 1902 he opened Bunbury's first auction room, in combination with his land and estate agency in Dixon's Buildings, opposite the Willows on the corner of Victoria and Stirling Streets.

When he came of school age, Charles attended the State Primary School in Arthur Street, at the foot of Stephen Street, where Thomas William Paisley was headmaster. Nora remembers Caroline's pride—and that of Master Charles—when he came home to boast 'Mr Paisley patted me on the head today and said, "You're a good boy, Mitchell!" '

Charles retained a great respect for his first headmaster, who remained in Bunbury until his death in 1928. Writing to his mother in 1917, Charles commented: 'Remember me to [Mr Paisley] when you see him. I have always a warm spot in me for him. I always thought him very fair and he never spared the cane, particularly with me, my tongue and spelling always brought me trouble.'[2]

One of Charles's morning and evening chores was to escort the house cow to and from its grazing on the swamp land to the south of William Spencer's property.[3] Spencer had one of the few orchards in Bunbury and it posed a challenge to every redblooded schoolboy. Stolen fruit always tasted the sweetest and our hero not only had a sweet tooth but as the Mitchell and Spencer properties practically adjoined, every opportunity to indulge it.

Spencer had imported a valuable orange tree which was his particular pride. It may have been a navel or late valencia type, although official records are to the effect that these types were not generally available till some years later. It first fruited in Charles's eleventh year with three lovely golden oranges which Spencer could hardly wait to sample. Unfortunately he delayed one day too long. There was a tremendous outcry, with Charles the number one suspect, and even the police were brought into it. In later life Charles never conceded that the fright and the scolding he received were worth the oranges. 'They tasted sour', he recalled. 'Even the cow wouldn't eat them and, anyhow, why should old Spencer make a fuss with our Jim married to their Clara.'

Spencer would never have been prepared to overlook the incident as 'within the family'. On the contrary, although the marriage of James and Clara on 17 September 1888 was a love-match, it never had the unreserved blessing of William Spencer. He was one of Bunbury's leading merchants, he owned possibly the best residence in the town, built on the best site, and he was also a town councillor with his eye on the mayoralty as a stepping stone to a political career. His daughter could do far better than to marry a bank clerk. The Bedford Mitchells said little, but thought a lot. Their turn would come.

Two events occurred in Charles's twelfth year (1891) which he often recounted. The first was when, during the school holidays, the hot and dusty time of the year following Christmas, he was entrusted to deliver a riding hack sold by his father to John C. Allnutt of 'The Grange', Bridgetown.[4] For a small boy this was a considerable journey (of 60 miles) over rough and usually unfrequented bush roads. He left at daybreak, armed with a great pile of sandwiches, and covered the route by mid-afternoon, having been instructed not to arrive with the horse in a lather of sweat, or looking jaded. Allnutt took delivery, thanked Charles, gave him a shilling and then left him to his own devices. Charles, displaying even then the sturdy independence which was to distinguish him all his life, walked and ran the 6 miles to Bridgetown. There was no coach till the following noon so he importuned the hotel keeper for meals and a bed against 'my Dad will pay the bill'. He dealt similarly with the coach driver next morning to get home to Bunbury. The shilling he had earned was his own, and not to be spent on his father's business.

The other event he recalled so well was the opening, on 12 March, of the Bunbury-Boyanup railway. This event was celebrated by the entire Mitchell family—with the rest of the town and district, reported to be some 800 in all—riding by train out to Boyanup. The official opening was held at noon and the celebrations which followed, sports and private picnics, occupied the whole afternoon. The Mitchell children chose to ride in an open truck and it is easy to understand the experience living on in the memory of the man who, as a child, rarely rode in anything faster than a sulky. It must have seemed something like a miracle for him to flash through the green south-west countryside behind the puffing monster of an engine. The Mitchell children were, in fact, to be very well catered for in the matter of railway excitement.

Apparently so that the proposed station would be placed in the centre of the town, the railway line turned west at its present intersection with Cornwall Street, then north around the back of the Mitchells' place, to the terminal and goods sheds approximately where the Convent of Mercy stands today. The children enjoyed having a railway practically at the door, although for months the horses and stock took fright every time the trains blew their whistles, and the older people complained of the noise of the steam pump raising water to overhead tanks from the nearby bore. However, the excitement, inconvenience and noise lasted only until 1893. In that year Bunbury was linked to Perth by rail via Picton Junction and the station and shunting yards were transferred to their present position.

On 8 September His Excellency the Governor, Sir William Robinson, introduced by William Spencer (who had meantime achieved his ambition to become Mayor of Bunbury) and supported by the Premier, Sir John Forrest, K.C.M.G., declared the Perth-Bunbury railway line officially opened. It is well worth recording that whereas Collie River coal was used exclusively to bring the special train from Perth, engines on the Bunbury-Boyanup run were fired with wood.

Bunbury was all agog with the completion of the rail link. The potential of the nearby coal deposits and the promising development of the tin field discovered at Greenbushes five years earlier presaged a bright future for the town, and the speakers at the opening and official dinner lost no opportunity of emphasizing the new and glowing era into which Western Australia generally, and Bunbury in particular, were about to enter.

On at least one other recorded occasion Caroline was proud of Charles. A Mrs Nankervis called to thank her for Charles's having defended her young son from some bigger boys who were bullying him. No-one then recognized the signs, but what activated Charles was not the love of fighting for fighting's sake, but the instinct to protect the underdog, for whom he battled the rest of his active life—mostly unasked and seldom thanked.

Caroline carried her deep and unabiding hatred of strong drink to the presidency of the local Temperance League. One evening as she was holding forth in the church hall young Charles broke the meeting up—mainly in merriment—by calling out through a window: 'What about the pub you own at Donnybrook?' He himself could see no harm in drink: his father and his father's friends all drank with no more apparent result than a pleasant atmosphere of hospitality and joviality. Although he was most likely put up to this prank by his elder brother Will, his mother, normally indulgent, failed to appreciate what was to his developing character merely a joke. Enough was enough: young Charles needed discipline, and it would be most satisfactorily applied at brother Harry's old school: High School in Perth.

'Arrange for him to be booked in next year!' was the dictum she issued, and it was obeyed. Perhaps it is not correct to surmise that the childish escapade which so upset his mother was alone the reason for his being banished to boarding school in Perth: the Mitchells were ambitious for their children and, much as they loved their pleasant little Bunbury, it was at that time unable to supply the educational polish they could afford for their boys. In all probability Charles would have been sent to Perth willy-nilly, but whatever the reason, he now severed the close connections and relatively sheltered life of his childhood and, for the first time, moved into that wide world in which he was to make his mark so decisively in the coming years.

1. Children born to William and Caroline: James, born 27 April 1866, married Clara Robinson Spencer 17 September 1888, died 26 July 1951; Kate, b. 15 April 1867, spinster, d. 28 December 1960; William Bedford, b. 6 September 1870, m. Fanny Priscilla Pearse 12 May 1898, d. 25 March 1928; Henry Arthur, b. 6 September 1870, m. Minna Sophia Thomson 6 January 1898, d. 5 October 1940; Laura, b. 6 February 1872, d. 15 February 1872; Janet, b. 9 February 1873, d. 17 December 1873; Elinor, b. 22 August 1874, d. 25 October 1874; Mabel Edith, b. 13 September 1875, m. Stephen Henry Fry 9 August 1905, d. 24 July 1956; Amy, b. 24 February 1877, spinster, d. 31 July 1946; Charles Douglas, b. 20 July 1879, m. Ida Marian Rossiter 13 June 1905, d. 19 February 1967; Nora, b. 3 November 1880, m. William John Rae, widowed 27 March 1917; Rose, b. 25 March 1882, spinster, d. 4 November 1901; John Bedford, b. 4 December 1884, d. 27 July 1885; Marion, b. 28 April 1886, m. Harry Chatterton Nankervis 6 July 1909, d. Adelaide 3 March 1958; Thomas Bedford, b. 24 April 1888, m. Eileen Carse 1932, divorced 16 November 1938, m. Doris Eileen Coxon 28 March 1942, d. Melbourne 25 February 1944.
2. Letter from Charles Mitchell to his mother, 23 June 1917, in the possession of Mrs Nora Rae.
3. William Spencer, merchant of Bunbury, purchased from George Eliot the first home built in Bunbury on an area of 19 acres 20 perches, 'being part of Leschenault Lot 26', the original grant to Sir James Stirling, subject of Certificate of Title Vol. IX, Fol. 392. The property extended from Stirling Street in the north to present-day Sampson Road in the south. The eastern boundary is today's Spencer Street (then called Vasse Road) and the western boundary approximately the prolongation of the western side of Victoria Street. The residence was on 'Bury Hill', the site of today's St John of God Hospital, while his shop was in Victoria Street, opposite the original Church of England cathedral.

   There were ten surviving children of his marriage to Hannah Properjohn: Hannah, Frances, Ellen, Clara, George, Charles, Augusta, Llewellyn, Victor and Zoe. Spencer was Mayor of Bunbury, 1891-92; he retired from the Bunbury Municipal Council in 1897; and he was a member of the Legislative Council July 1896–July 1901, being the member for South-West Province till his death 23 July 1901.
4. Now held by the estate of Herbert Coleman Davies, approximately 3,100 acres, 6 miles south-west of Bridgetown.

4

Secondary School Days 1895-1896

Following in his brother Harry's footsteps, Charles was enrolled as a boarder at High School in February 1895.

The school, later known as Hale School, was situated at the northern end of St George's Terrace, with frontage also to George[1] and Hay Streets. It claimed continuity with the secondary school established by Bishop Hale in 1858, then occupying part of the present Cloisters.

From the rear of the headmaster's house, which faced the Terrace, a high stone wall with broken bottles embedded in the top ran north to enclose the school, dormitories and refectory. Entrance for the day pupils was from the oyster-shell footpath in George Street through a postern gate.

When Charles was enrolled there were approximately a hundred pupils of whom some twenty were boarders. John Barrett-Lennard, late of 'Mundathoola', near Williams, recalled that as new boys in 1896 Phil Fry of Bunbury and he were the last pair to march into the main hall for the first assembly. The opening words of the headmaster's address on that occasion were: 'I have much pleasure in announcing that the hundredth boy just walked into the school, and in honour of such an event there will be a half holiday tomorrow.'

The first of several distinguished headmasters, F. C. Faulkner, M.A. (Cantab.), occupied the position from 1890 to 1914. Other masters were E. W. (Paddy) Haynes (a former headmaster), Fannin Evans, J. Marychurch Jenkins and the sportsmaster, F. D. Stephenson.

There is no record of the academic achievements of the boy who was now known as 'Mitch', but he is remembered for two things.

He was the 'Ginger' Mitchell who fought a Homeric bare knuckle bout with one Thompson, greatly to the delight of the younger boys who normally were the butt of their seniors; and he was one of the six High School boys who volunteered for the Boer War.

John Barrett-Lennard remembered well his first evening in the dormitory. Phil Fry and he, of course, were allotted beds in the worst situation right at the far end of the long room. There the pair of them sat, petrified, in their pyjamas, while the bigger boys played their usual boisterous games. Suddenly, John recalls, a great red head popped up in front of them and a comfortingly strong voice said: 'I'll look after you two kids tonight, but after that you're on your own.'

The protection went beyond the night, however. Going into breakfast next morning one boy cuffed John and in reply to a: 'Leave him alone!' from Mitchell, 'offered Mitch out' as soon as breakfast was over. They accordingly adjourned to a small enclosure, between the main building and the high stone wall bordering the road, and set to. This bout should not be confused with the 'Homeric' battle, as it was soon stopped by the appearance of one of the housemaids, who belaboured Mitchell with a broom, calling him 'a pugnacious redheaded beggar'. John never accepted the housemaid's description. As far as he was concerned he never met 'a nicer spirited boy' in all his life.

The immediate 'unofficial playground', the very rough and sandy vacant land bounded by St George's Terrace, George, Hay and Milligan Streets, comprised the whole of the area with the exception of that part occupied by the school and two residences facing the Terrace and adjoining the school boundary on the east.[2] Football, cricket and athletics were conducted on the school oval, which was situated on the site later acquired and occupied until June 1967 by the Perth Observatory.[3]

Mitchell, of course, played football. Among his most treasured possessions was a photo of himself among his fellow members of the school's 1895 First XVIII coached by Stephenson and with either Norman or Eric Ferguson as captain; they were twins, and no-one can remember which played football.

The school colours were, and still are, the dark and light blue combination of Oxford and Cambridge, but the guernseys were many and varied: from lace-up fronts, broad and narrow horizontal stripes, to full colour and vertical stripes, plus three-quarter length trousers and short stockings. Certain of the players were identified in later years, in Mitchell's round hand, as 'Sandy' More, Billy Worth, W. J. Moll, 'Buzza' Marmion, Harry Yelverton, Clayton Timperley, George Clarke, Frank Alan, Charlie

Shenton, Reg Courtenay, Wilfred Lee-Steere and Claude Passmore. R. G. (Reg) Courtenay was eventually chief accountant for the Agricultural Bank from 1931 till his retirement in October 1944.

Reg, half-back and goalkeeper, recalls that Briggs's school[4] in Fremantle, at the time the only other private school in the colony, played both cricket and football against them. Each Wednesday afternoon, which then was the business houses' half holiday, the High School team played junior teams organized by local firms, and occasionally competed at Guildford against a team composed of youthful employees from Ferguson's vineyards. Reg recalls, also, that the Annual Sports Day was first held at the W.A.C.A. in 1896.

George Field, the master who succeeded Paddy Haynes, came from England and was a keen cricketer. He endeavoured to have soccer made the school's winter sport. A referendum was held and resulted in a win for 'Aussie Rules': it was probably unique in college history and indicated the democratic spirit of the young, bustling colony, which had only recently been granted self-government.

Swimming, boating and paper chases ranked next in popularity, swimming sports being held off Perth Spit, a sandy bank in the middle of the river, approximately opposite Mill Street of today. Paper chases were routed through Kings Park into the fully timbered wilderness of Nedlands, where the 'hares' could invariably lose the 'hounds'.

Paddy Haynes exercised an ingenious form of punishment: for the first misdemeanour, a boy received 'fifty squares', which meant that he had to rule a sheet of paper into fifty squares and then in the appropriate space, square each number from one to fifty. For a second offence the boy squared each number from fifty-one to 100, and so-on. Paddy considered lines a waste of time; apparently mathematics were his love.

The outstanding scholar of this particular era was Charles 'Buck' Ferguson, while the head boy was Frank Parker, who later served as Colonel Parker A.I.F. in the 1914-18 War and was killed in action in Egypt. H. S. D. Rowe and Ernest Parker were the college's star cricketers and both later represented Western Australia.

The drill master was Colonel Angelo, Commandant of the Western Australian forces, a strict disciplinarian who had served under General Roberts in India. It is possible that his training and influence explain the acceptance of under-age Haleians into the colony's first expeditionary force. Lieutenant Tony Forrest, aged seventeen, was one of the six High School boys who enlisted for service in the Boer War.

Harry, Charles's elder brother by nine years, was the younger boy's idol. A painstaking and ambitious student who had won one of the first scholarships made available to High School, Harry bore out his early promise, delighting his family by returning home with a trunk full of prizes. He was not only a good student but, like most youngsters in those times of 'make-do' and independent initiative, he was also a hardy customer. During his time at High School, travel between Perth and Bunbury was by stage-coach, as the railway from Perth to Picton Junction was not opened until 1893, although the Bunbury-Boyanup line through Picton began running in 1891. On the occasion of the break-up of school for the first Christmas vacation (in 1886) Harry, with his friends, Will Brockman of Warren Bridge and Archie Bunbury, having spent their coach fare, walked home—a matter of a mere 100 miles.

During Charles's term at High School his beloved elder brother had qualified as a surveyor, was granted a limited licence on 9 January 1895[5] and entered into partnership with Newton Moore of Bunbury.[6] The partnership was short lived.

Opportunities and experience abounded in the recently discovered East Murchison goldfields, and to take advantage of them, Harry joined A. W. Newman's survey party at Southern Cross in early September 1895. While with this party he was granted (on 7 February 1896) a licence under the Transfer of Land Act to carry out Titles Office surveys. On the death of Newman at Cue in May of that year Harry, now with an unrestricted licence, formed his own party. In June he accepted his first contract from the Lands Department, and set out for Lawlers, some 230 miles north of Kalgoorlie.

Charles turned seventeen in 1896, and he left school in August of that year. School irked him. At best, he competed equally with his contemporaries only at sport. In the academic subjects he lagged far behind. The headmaster, F. C. Faulkner, sent for him just before the mid-winter holidays, told him how disappointed he was in him, how he felt Charles was wasting his time and his people's money, and that he was writing to his father to that effect. To the headmaster there appeared no reason for Charles to return after the term holiday.

First XVIII, High School, 1895

A. W. Newman's survey party, Southern Cross, 1895

With such a report Charles did not relish facing his father, and the thought of joining Harry's survey team occurred to him. At least it would offer him an escape from returning to Bunbury until the storm had blown over. Here was adventure and the outdoor life, and Charles was certain that Harry would welcome him. It is uncertain whether he made prior contact with Harry, but by the time the family had received a letter telling them he had left school, he had caught the train to Boorabbin, which was the end of the line. The 80-mile extension from Boorabbin to Kalgoorlie was not opened until January 1897. Cobb and Co's coach took him the 57 miles to Coolgardie and a horse team, carrying supplies, deposited him at Lawlers. His boyhood and schooldays were behind him, and he had entered simultaneously into manhood and the exhilarating tide of affairs he was to ride for many a day.

1. Renamed St George's Place on 29 November 1935.
2. Anthony O'Grady Lefroy occupied Lots 6 and 7 of Lot H12 and Richard Adolphus Sholl, Postmaster-General of Western Australia, resided on Lot 1 of Lot H11.
3. The school governors were granted as 'Tenants at Will', under advice from the Under Secretary for Lands, dated 4 December 1897, an area in lieu (approximately 4 acres) within Kings Park Reserve.
4. Briggs's school (Fremantle Grammar School), situated at the top of High Street, Fremantle, was owned by Sir Henry Briggs, a member of the Legislative Council and later its President. He was knighted in 1916.

   The residential portion for boarders and staff was the delicensed Rose and Crown Hotel on the site now occupied by the Oriana Theatre. Briggs's later became a girls' school—Girton College—and today the building is a Church of the Latter Day Saints.
5. W.A. Licence No. 14 was issued 9 January 1895, which entitled Mitchell to work for the Lands Department only. On 7 February 1896 he was granted a licence under the Transfer of Land Act to carry out Titles Office surveys; this was subsequently registered under 1909-1958 Act as No. 157 on 11 January 1911. The following extract from a letter written by H. A. Mitchell to the Commissioner of Titles on 9 January 1895 is of interest: 'In reply to your correspondence of the 27th December, 1895, referring to my application to be specially licensed to practise surveying under the Transfer of Land Act, 1893, I beg to state that I have had a great deal of experience in the field at all classes of work connected with land surveying. I worked three years with Mr. G. R. Turner, five months

with Mr. C. E. May, two months with Mr. C. M. Denny, nine months with Mr. A. W. Newman, and with Mr. N. H. McDonald and H. I. Farrell for a few months.'

6. Newton Moore was Premier of Western Australia for four years beginning in 1906. He had the C.M.G. conferred in 1908, and was knighted K.C.M.G. in 1910.

5

Surveying and Tin Mining 1896-1899

The goldfield to which Charles so laboriously made his way in August 1896 was one of many which flourished then waned in Western Australia about the turn of the century.

Following the discovery of gold in 1885 at the junction of the Elvire and Panton Rivers in the Kimberleys, and the subsequent alluvial finds on the Margaret and Ord Rivers, the Kimberley goldfield was proclaimed in 1886 and Western Australia experienced its first gold rush. Production was not spectacular but sufficient to maintain interest. 'Gold is where you find it': the signs were encouraging and prospectors roamed far and wide. Southern Cross was proclaimed a goldfield in October 1888, with the Pilbara field following in 1889. When Arthur Bayley and John Ford discovered gold in Coolgardie in 1892, and were followed shortly afterwards by Paddy Hannan with his find in Kalgoorlie, Western Australia's future seemed assured. The colonial government pledged all its resources, in priority to all other demands, towards the public servicing of gold production.

At Darlot and Lawlers, gold was first found in 1894, and on 28 June 1895 the East Murchison goldfield was proclaimed and placed under the control of Warden A. G. Clifton.

Lawlers was the main settlement, and Harry's first commission from the Lands Department was to survey the townsite reserve. He allotted an area of 500 acres for this purpose and, fitting it in with the existing scattered settlement, laid out Wallace, Parke, Turner, Sutton, Clifton and Newman Streets, all 150 links wide and named in honour of prominent local citizens and early 'diggers'. The next assignment was to locate and survey the Lawlers cemetery site. The local 'boot hill' was no longer socially acceptable; the populace were entitled to respectability, consecrated ground and tombstones. They could now afford them. However, Lawlers did not develop as planned, for its peak population of 542 in 1901 had declined to 100 by 1911.

The survey of the township had been accomplished before Charles arrived in late August and, as he had hoped, had been received with open arms by his elder brother. He arrived, in fact, just in time to take part, as 'horse boy' and assistant chainman, in running a traverse from Lawlers to Mt Sir Samuel, some 25 miles to the north, and there to locate and lay out a townsite which was eventually selected adjoining Mining Lease 180. It had been recommended that this town be called Herberton after a prospector of the locality, but the recommendation was not accepted and it was later named Newmanton.[1]

Returning to Lawlers in September, the party—usually about four or five men transported by horse or camel in drays, carts and buggies—next set out to run a traverse 50 miles to the east of Lake Darlot, named after L. H. Darlot, a local squatter. Here the job was to locate and lay out a township to be called 'Woodarie' after the Aboriginal name for the locality.[2] For some reason not recorded, this name was altered to 'Woodara' and was proclaimed on 14 January 1898.

The survey parties used horse teams to draw sulkies and drays, and camels to carry heavy packs. Young Mitchell could look after the horses, but the camels, which were hired, were tended by the Afghan owners.

Christmas 1896 was spent at Darlot, and was long remembered by Charles for the heat (never under 100°), flies, dust and a painful complaint from drinking heavily mineralized water. Dehydration resulted in a high concentration of solids in the body, and drinking the mineralized water often produced disquieting urinary symptoms. This was the first and only time the budding chainman wished he had stayed at school.

The gold strikes in and around Newmanton and Woodara did not develop to the extent that had been hoped and in effect these towns, like many other similar little settlements, were stillborn.[3] Small shows in the Darlot district continued to produce, however, and a state battery, operating full time, is still maintained.

Harry had entered into a working partnership with W. Charlesworth Smith, a qualified surveyor and civil engineer. The latter and his brother, W. Staniforth Smith, had migrated to Western Australia from Victoria with a joint capital of £50, all their father could spare from the land boom and bank crash of 1894. The partnership was successful, each partner respecting and trusting the other. Their main income was derived from the survey of mining leases and laying out the plot for underground workings, for which work they were paid by the individual mining companies. Warden Clifton went out of his way to help the partners, acting as their agent, collecting and following up accounts. His help and friendship was invaluable, if only because distance and lack of communications rendered impossible the normal business procedures and arrangements for credit, as they are known today.

The mining surveys took the partners in all directions from their main camp at Lawlers and, of course, young Charles went with one or the other team. Early in 1897, when returning from the Yundamindera district, the party camped at Cardinia Creek, south-east of Malcolm. The following morning Charles found that a number of his recently acquired charges were missing, and a quick search revealed that the horses had started eastwards along the telegraph line leading to Murrin Murrin. Their straying was a reflection on Charles's capacity as horse boy, and he begged permission to go after them.

'Alright', said Harry. 'Take some tucker and water-bags and catch us up as soon as you can.'

Tracking was easy for some 10 miles, until the horses veered slightly south, heading for the well-grassed country surrounding Kilkenny Creek between Murrin Murrin and Eulaminna, where the tracks were no longer easily discernible, and Charles had to guess which hollow they were likely to follow for feed, and to keep his eyes open for fresh traces to confirm his guesses. With many a blind lead, he tracked them as far north as Mt Morgans. They had milled around there on the edge of gibber desert looking for feed, and he caught up with them. The search had taken six days and covered nearly a hundred miles. It was seven days and 200 miles later before Charles made contact with the main party. Although he had only just turned eighteen, the youngster made nothing of the trip, accomplished entirely on his own—a 300-mile ride on horseback across Western Australia's near deserts was part of the job. Perhaps it was the writing home of these exploits that later prompted his mother to insist on his return to Bunbury.

Towards the middle of 1897 Harry and his party, with the exception of Charles, returned to Perth. Charles, either by choice or direction, transferred to Charlesworth Smith's party, and their first work on behalf of the Lands Department appears to have been their July 1897 traverse from Lawlers to Wiluna to select and lay out a townsite reserve.

Prior to 1896, when gold was first discovered there, the district was commonly known as 'Lake Way' and the township as 'Weeloona', a native word accepted as meaning 'Windy Place'. The spelling became abbreviated, probably by phonetic adaption, and the township of Wiluna was proclaimed on 4 April 1898. There is a suggestion that the name 'Weeloona' came from the call made by curlews around Lake Way and Lake Violet. Ernestine Hill wrote to this effect in the *West Australian* (21 June 1952): 'Weeloo! the curlews are crying around the shores of Lake Way and Lake Violet.'

Wiluna rose to fame in 1897 when both alluvial and oxidized ores were discovered, and particularly when a prospector named Martin, on 29 December 1897, found the 463oz nugget 'The Monarch' which up to that time was the largest found in Western Australia. W. Charlesworth Smith wrote to Harry on 2 March 1898 conveying news of this find: 'Bob Martin found a 450oz nugget before Christmas last year and kept it quiet for months'[4] —the story having been that Percy Martin buried the nugget and refused to disclose its whereabouts until his brother Bob, for whom he had been working, gave him £500.

There was ample survey work offering in the Lake Way district and it was not until December that the party travelled to Peak Hill. Christmas Day 1897 was celebrated in Cue.

Charlesworth Smith referred to Charles's departure in his letter of 12 March, to his partner from Lawlers:

> Charles left here for Bunbury last Thursday per coach to Menzies then by rail. The through fare by coach and rail to Perth is now £7. He was going to Lake Way with me where there are two or three surveys on hand. However, the day before we were to start for Lake Way, I got a wire from Mr. Anderson to say Edward Gilliband had arrived in Cue and as Charles' mother seemed anxious that he should come down for a change we thought it would be just as well for him to go straight down and for me to wait for Friday's coach from Magnet for Edward who arrived tonight. I can hardly say that I am very sorry to lose Charlie. As he did not think he would follow up surveying and as one of his brothers was talking of starting an auctioneering and agency business he thought his brother would be able to find something for him to do and which would be more in his line—so I did not press him to stay, as this game is not up to much unless a man makes up his mind to master it.[5]

In the face of this letter it is difficult to determine who was the more responsible for Charles's returning to Bunbury, he or his mother. It could have been merely that he himself was fed up after having been more than two years away from home in a calling he did not, basically, wish to follow. It is possible on the other hand that Caroline, influenced by between-the-lines reading of his letters, suggested that he come home, at least for a while. Moreover, with James and Will in the banking profession and Tom still in primary school, there was no son at home to assist her husband.

Whatever the cause for his return home, his stay was not for long as tempting employment was scarce in Bunbury. Instead of returning to the East Murchison, Harry had accepted contracts from the Lands Department which took him into the Kent district, and in November 1897 his party, including Charles, was engaged in defining farming locations in the Bremer Bay and Gardiner River areas.

They were very busy, and expected to take the minimum break for the Christmas 1897 and New Year holidays. Harry had decided on a wedding date, however, so prolonged his stay until his marriage to Minna Sophie Thomson, at St Paul's Pro-Cathedral, Bunbury, on 26 January 1898. The Thomson family had pioneered the Brookhampton district, and their old home survived until 1956, when it was destroyed by fire. Although the bride's mother was Emma, daughter of John Septimus Roe, Western Australia's first Surveyor-General, the relationship apparently did not carry any influence. The family recalls that the Surveyor-General, H. F. Johnston, mulcted Harry £10 for marrying without his prior permission. However, it is thought that really the fine was for absenting himself unduly, without the Surveyor-General's knowledge and consent.

Work was still pressing, and Minna left with Harry for a honeymoon at Bremer Bay. She lived in a tent, possibly serving as camp cook to the team when they returned each evening. Bremer Bay must have had some romantic appeal, as apparently there and then it was resolved that their branch of the family, to distinguish it at least from the numerous non-'Bedford Mitchell' families (and there seemed then to be more Mitchells than Smiths in Western Australia), should be known as the Bremer Mitchells. And so it is: their children Marjorie (Mrs Staniforth Smith), James, of 'Uda Bremer', Donnybrook, Guy, of 'Sylvan Loch', North Dinninup, and Minna (Mrs Tom Brockman), and their grandchildren, all carry 'Bremer' as a second Christian name.

Harry's acceptance of surveying contracts in the lower South-West evidently had the approval of his partner. A letter from

Charlesworth Smith, dated 2 March 1899,[6] indicated planning on behalf of the partnership but no subsequent references to any future joint efforts can be found.[7]

Charles did not go on the second trip to Bremer Bay, and after a holiday (in his opinion a well-deserved one), joined Newton Moore's survey team, the reason in this case being that Moore's team never moved further from Bunbury than Harvey or Collie, except, as Charles related in later years, when the party ran a line from Augusta to Walpole. Charles was unimpressed with the country traversed. The soils were either sand or swamp, or were covered with exceedingly thick scrub. When they were not bogged, they were either wading or chopping their way; it is possible that this one experience above all others turned him to the wide open spaces of the wheatbelt.

Charles enjoyed his few months with Newton Moore, who was very kind to him. Kindness notwithstanding, he rejoined Harry who, in order to be near his bride, had accepted contract work for the Lands Department in the Wellington and adjoining districts. In July 1898, with Charles as chainman, they surveyed Wellington Locations 914, 988, 989 and 992, 3 miles north-west of Donnybrook.

The next three months saw them laying out 'suburban for cultivation' lots (5 acres) at Donnybrook and Boyanup, completing the design of the Preston agricultural area, and defining various farming locations in the Ferguson River district. During December 1898 they defined Locations 813, 814 and 856 in the Nelson area, 3 miles north of Boyup Brook.

Harry and his wife had made their first home at Donnybrook, in a two-roomed cottage which Minna had had built for herself when she left her childhood home at 'Old Brookhampton' following her brother Jack's marriage and the decline of the need for her services as housekeeper. For some unrecorded reason the Donnybrook cottage was always known as 'The Folly'.

Harry later purchased from Charles Raymond Trigwell a small farming property (Wellington Location 455) across the river from the Folly, adjoining a property of William Bedford's known as 'Kellys'.

Charles could thus always fill in his time gainfully, assisting either at farming or surveying. Besides, he was not far from Bunbury and home. Maturing in the shade of his elder brother, listening and learning, he was now a competent survey hand and chainman, capable also of relieving as cook or teamster. Harry was meticulous in his work, and under his tutelage Charles learned the importance of the correlation of seeming trivia, adopting such careful practices with confidence and humility in his later life. But apparently either he had had enough of surveying, in which progress required his re-application to academic subjects (the thought of which was unbearable), or he was looking for other fields to conquer. Suddenly an opening presented itself.

In or about September 1899 Harry and Charles arrived in Greenbushes to survey several mineral leases for the Mines Department, which at that time did not have its own surveyors, but passed its work to the Lands Department.

The Greenbushes Mineral Field was discovered in 1888 by David William Stinton, a local resident.[8] As gazetted on 22 March 1907, it was of some 2,500 acres. Tin could be found in scattered alluvial patches or deep lodes at an average depth of 100 feet.

Until about 1910, when steam-powered floating dredges with portable nozzles and sand pumps similarly powered were introduced, all sluicing operations were carried out manually. A four-head stamp battery was built at about the same time.

The system employed was to get out as much tin-bearing material (known as 'wash') as possible during the summer months, dump it below the site of a dam constructed across a creek or depression which would then fill with the first winter rains.

As soon as sufficient water had been stored, sluicing would commence. A wooden box or strake (an inclined trough) would be built below, but as close to the head of the dam as possible. The length of the strake was anything from 8 to 10 feet to 40 and 50 feet, depending on whether it was a one-man show or operated by a company. The soil was shovelled into the box and water was then fed, with force if possible, into the box to carry the material down the strake, across which were built baffles of a heavy ribbed cloth. These formed a number of compartments in which, because of its weight, any tin would fall to be removed at the end of the day. The problem was to find tin and then have water available at the site.

It appears that when Harry and Charles came to define the boundaries of one lease it was found to have been considerably 'overpegged' by the prospector, beyond the 48-acre limit.[9] This left an area of 'no-man's land' and one can surmise that young Charles saw an opportunity where, for the investment of ten shillings for a prospecting lease, he could get into the tin business.

There is no record of his being issued with a Miner's Right at either Greenbushes or Bridgetown, and it is presumed that like every other 'goldfieldite', he had carried one with him throughout the whole of his East Murchison surveying venture, never missing an opportunity of either 'napping' or 'specking'.

The field was a highly profitable one. For the ten years 1891-1901 the yield was 2,624.61 tons of tin, valued at £68,073. According to Mitchell, however, his contribution was no more than sweat and blisters. There is no record of his ever having been in production, by applying for water rights or a mining lease, and besides, there had been little time.

Within two months of his arrival at Greenbushes came the declaration of war against the Boers in South Africa. That was the finish of tin mining for young Mitchell. He wrote home immediately, asking his family to use all their influence to get him into the volunteer force. The fact that he was not twenty-one years of age did not worry him. He made his way to Donnybrook and caught the next train for Perth, stopping off at Bunbury just long enough to enjoin his family 'to go to work on' Newton Moore, who held senior rank in the Western Australian Volunteer Corps, who knew the family well, and was brother Harry's erstwhile partner. The boy who had so gamely delivered his father's horse more than 60 miles from home in the sparsely populated South-West, and the young man who had so unconcernedly traversed 300 miles of the even more sparsely populated and treacherous semi-desert of the Eastern Goldfields country, was about to embark on a much more hazardous journey.

In a curious way it was to teach him, through his country's enemies, another of the lessons which were to add to his total commitment to the welfare of his fellow men.

1. Following the usual custom of recording local identities, the streets were named McBean, Barnes, Clarence, Anderson, Furman, O'Grady, Herbert, Bizant, Abbot and Wilson.
2. Mitchell named the streets Rogers, Taylor, Darlot, Yibeah, Lockhart, Jannet and Benson, but Darlot was the only name preserved in a later Mines Department plan.
3. The Mining Act permitted holders of Miner's Rights to peg areas, outside of townsite reserves, for miners' residences, business sites and miners' homestead leases (farms). Thus many miners had made arrangements for their personal convenience prior to the establish-

ment of a town, which, in effect, followed exploration and proven development and provided for the community's social servicing requirements only when population density made it necessary. After World War I, legislation provided for miners' homestead leases being brought under the Land Act. The privilege has since been withdrawn.

4. Extract from a letterbook held by Mrs Marjorie Staniforth Smith, Bridgetown.
5. *Ibid.*
6. *Ibid.*
7. The partnership had purchased building sites for branch offices at both Lake Way and Darlot. Charlesworth Smith advised, 'bought two good chairs from Solley when his wife left, for 12s', as office equipment, and then continued to express his confidence in the future of these districts: 'With reference to the Lake Way blocks I think they will be very valuable in time as that field has every indication of permanence. There is already two small batteries there, one 5 head and the other 3 head—the former on Cook's show and the other owned by W.A. Goldfield Coy. for testing small parcels of stone . . . There is also a ten head on the road and which is going to crush for the public and I hear there is another ten head going to be put up, so that if they get 28 stamps going there this year it will make a great difference to the look of the place.'

   Charlesworth Smith here leaves our story, but his brother, Staniforth Smith, re-entered it in 1928 when, on 28 April, he married Harry's daughter Marjorie at St George's Cathedral, Perth.
8. Stinton was granted three discovery claims, each of 20 acres, and in 1891 was paid a reward of £250. The reward was the result of a petition, signed by 137 people, addressed to the Premier, Sir John Forrest, through Sir James George Lee-Steere, member for Nelson in Western Australia's first parliament. In 1908 the people of Greenbushes and adjoining districts presented a further petition to the Premier, Sir Newton Moore, to increase Stinton's reward from £250 to £400, but without success. Stinton did not continue in the industry and in 1891 sold his lease to A. Knox-Brown.
9. Whilst gold leases were restricted to 24 acres, both mineral and prospecting leases had a maximum of 48 acres.

6

The South African War

Great Britain declared that a state of war existed between herself and the Boers on 11 October 1899. The Western Australian government sanctioned the raising of the first company of mounted infantry by notice in the *Government Gazette,* 13 October, and of the second contingent on 20 December. Amongst hundreds of volunteers who reported at Karrakatta on the morning of the twenty-first was young Charles Mitchell.

Family pressure had triumphed; Newton Moore was at the Recruiting Office. A sergeant put his head around the corner and called out: 'Charlie Mitchell to report to the office immediately.' Volunteers with nothing better to do took up the call and repeated it with great enthusiasm. Anyhow, nobody knew Mitchell. Who was he? Charles, covered with embarrassment, virtually slunk into the office, was examined and accepted; apparently his being in his twenty-first year was good enough for the age requirement. Besides, there was Newton Moore's assurance of family consent. He walked out as Private Charles Douglas Mitchell, Regimental Number 89, 2nd Contingent Western Australian Mounted Infantry (W.A.M.I.). He had described his previous profession as that of prospector.

Training commenced forthwith. There was so much to do and it was imperative that as soon as possible the contingents arrive in South Africa where their training would be completed.

Karrakatta had been established officially in 1885 as an artillery depot, and in fact two seven-inch guns had been stored there as early as 1861. Apparently, however, its purpose was primarily for Volunteer Cadet Corps training. An administrative headquarters was built in 1896 and the location of the rifle range was moved from Mount Eliza to Karrakatta in the same year.

What excitement reigned in the colony! The fledgling of Empire was at war, raising and financing its own contingent; and how dashing its men looked in uniform!

Charles's sisters, Mabel and Nora, were not going to miss out on any of the fun. They took the opportunity to come to Perth and stay with Mr and Mrs Henry C. Prinsep. Henry had joined the Public Service in 1874[1] and they now lived at 292 Hay Street —next to the Perth branch of the Royal Mint. Their picturesque home, known as 'The Olives', was hidden by shrubs and trees carefully planned and tended. It stood until replaced by the C. S. Harper Hall, headquarters of the Western Australian Adult Deaf and Dumb Society (Inc.) in 1958

Taking afternoon tea down to Karrakatta was the 'recognized thing'. True, there was an awfully dirty old train to travel in, and acres of sand to trudge through, but there were trees to sit under in the camp area, with seats provided for those who wished to watch the last of the day's training. How lovely the girls looked in their long white dresses, long sleeves and primly designed collars! Sailor-style hats—wide of brim and decorated with patriotic red, white and blue ribbons—were the rule. Black stockings and shoes almost invariably completed the ensemble.

Western Australia gave full support to the mother country, raising in all six contingents. Contingents 1, 2 and 3 of the Western Australian Mounted Infantry were of one-company strength, numbers 4, 5 and 6 of two-company strength, totalling, with reinforcements, sixty-four officers and 1,167 other ranks together with 1,179 horses. The nominal roll for a company comprised one major, one captain, three lieutenants, one captain medical officer, ninety-six other ranks. (Personnel of the 2nd Contingent, however, totalled 103 officers and men.)

Daily pay rates were: private 4s 6d, lance corporal 5s, corporal 7s, sergeants 8s. These rates included 2s 6d which was retained as deferred pay (forfeited if dismissed or dishonourably discharged or if a deserter). There were also issues of clothing and accessories.[2]

The colonial government of Western Australia had to meet the cost of these contingents from its own resources. The United Kingdom authorities eventually agreed to subsidize pay and field expenses, but details are not known.

The 2nd Contingent W.A.M.I. sailed from Fremantle in McIlwraith McEacharn's recently constructed 5,455-ton ship the S.S. *Surrey* on 3 February 1900, in company with similarly constituted contingents from New South Wales and South Australia.

The Western Australians' Commanding Officer, with rank of Major for six months, was in London at the outbreak of the War

—Captain Henry L. Pilkington, one-time officer in the Western Australian Volunteers.[3] He sailed from Southampton on 6 January 1900 in R.M.S. *Carisbrooke* to join his contingent at Cape Town.

The Western Australian contingent broke camp on Thursday 1 February, and that night saw the beginning of the experiences they were to undergo frequently in South Africa. Tents had been packed, as also had been all blankets except those which served as saddle cloths. The men, excited at the challenge, cheerfully chose suitable trees under which to pass the night, wrapped only in their saddle blankets. At 6.30 next morning they marched out of Karrakatta en route to Fremantle wharf. Cheered by spectators who intermittently lined the route, they arrived alongside the ship at 8 a.m. They had little difficulty embarking their horses, which were quite tractable after the somewhat tiring march.

Then came a memorable event, a march through Fremantle in the company of the New South Wales and South Australian contingents. Fremantle was the centre of attraction for some 30,000 spectators who had gathered to see, for the first time in Western Australia's history, the soldiers of their own and other colonies marching through the streets to receive the hospitality of the townspeople.

The mayors and councillors of Fremantle, East Fremantle and North Fremantle had entered so heartily into the work of preparation that they could not be denied the honour of entertaining a force of so decidedly federal a character. A sports meeting at Fremantle Oval[4] was followed by a banquet, at which members of the contingent were addressed by the Governor, Sir Gerald Smith, and Premier Sir John Forrest's Deputy.

On Saturday morning, the New South Wales and South Australian contingents went by train to Perth for a march through the city, followed by a reception at the Town Hall. The Western Australians were given the day for farewells to their wives and sweethearts, but only on the wharf, to which the ladies had been admitted by special pass.

On the voyage to Africa young Charles made an alarming discovery. He was no sailor and was sea-sick every day of the trip. To add to his discomfiture he seemed to be on perpetual stable duty.

The *Surrey* arrived in Table Bay on 24 February and the three contingents disembarked the following day at Cape Town. In

March the 2nd Contingent proceeded by rail and road to the Prieska district, some 53 miles north-east of Cape Town, there to join the troops engaged in quelling some organized disaffection. The rising having been put down, a part of the Western Australian force continued to operate in the district during part of April.

It has been assumed that Private Mitchell took part in these operations and from there was sent further north en route to Bloemfontein. He wrote to his family on 30 April:

> Bloemfontein. We arrived here yesterday morning from Springfontein on our way to Pretoria. We have been having a rough time of it lately, getting up at 1 or 2 a.m. and having very little sleep in the day time.
>
> We got our mail yesterday and some papers which Dr. Williams[5] kindly sent us, and enjoyed them very much.
>
> We met some of the first contingent yesterday, and were all very anxious to hear what they had to say. They told us all about the war, and the way they fought. We are all anxious to have a cut in. I don't think the war will last more than two months once we get into Pretoria, which we expect will be a hard fight. We are all in the best of health. This is a big camp. There seems to be thousands of tents. Eighty thousand men are here, so you can imagine what it is like! This is a very large town. I think it is the best place we have been in since we came over here.
>
> The horses are splendid, I have never seen anything like them; thousands of them and big upstanding horses at that. They would bring £30 to £60 a piece. Some of the Officers have beautiful horses. I have never seen anything like them in Australia. The Canadian horses are very good too. We have been escorting about 80 bullock wagons and 2,000 bullocks.
>
> We are all writing today although we have not much time for it. We always have something to do. We are expecting letters today. Lieut. Castilla[6] is the most popular man in the Contingent.[7]

While at Bloemfontein, the 2nd Contingent was attached to General Pole-Carew's 11th Division and took part in the May advance to Pretoria. They served with this division from 1 May to the end of October, with the exception of a few weeks about the end of June and beginning of July, when they were detached to assist on the lines of communication in the Orange River Colony. The 2nd saw action at Bradfort, 3 May; Vet River, 4th; Zand River, 9th and 10th; Vaal, 26th; Johannesburg, 28th, 29th, 30th; 6-Mile Spruit, 4 June; Silverton, 8th; Diamond Hill, 11th and 12th; and Belfast, 27 August.

Diamond Hill was a particularly heavy engagement and the Commander-in-Chief, Lord Roberts, in his telegram of 16 June reporting the action, gave praise to the Western Australians: 'Botha's army has retired, believed to Middleburg. His rearguard was surprised and thoroughly routed by Ian Hamilton's Mounted Infantry, chiefly Western Australians, and the 6th Battalion.'

On 16 July a party of the 2nd Contingent W.A.M.I. on the left of General Pole-Carew's position east of Pretoria, successfully repelled a Boer attack at Pienaar's Port. The 2nd took part, under General Pole-Carew, in the advance from Pretoria to Komati Poort, starting about 23 July. They had fighting in the Belfast district on 27 August, and frequently thereafter.

In his telegram of 13 September Lord Roberts remarked that the Western Australians had been scouting in front of Pole-Carew's division. They were said to have been the first to discover much railway material and other valuable stores near the Poort, and were present at a review on a large scale held there on 28 September. After their stay in that neighbourhood, they were employed about Machadodorp, throughout October, where they suffered some casualties.[8]

Mitchell's reminiscences of the war were few, but there are two which he delighted to recount.

Early in the campaign, while returning from a routine scouting trip of platoon strength, he and his companions passed an orange orchard. The heat and dust were stifling and their mouths watered at the sight of the luscious golden fruit. As soon as they were dismissed, Mitchell and two companions made their way back to the orchard. They had just filled their saddle bags when they were shot at by a considerable party of Boers. There was nothing to do but run for it, and in the opposite direction. The chase was long and the fugitives were saved only by darkness closing in. They then lost their pursuers, but their whereabouts was the next problem. It took three days to find their way back to camp, with oranges as their only diet. 'They never tasted the same, afterwards!' Mitchell always contended.

When the British forces entered Pretoria, the 2nd Contingent halted in the centre of the market square. There on the masthead of the town hall was the flag of the Boer Republic. Handing his reins to the nearest trooper, Mitchell shinned up the pole and souvenired the flag. He swears he gave it and other souvenirs to Lieutenant Henry Clarkson, who came home on the same ship, for custody and also to ensure they were allowed on the ship. That was the last he saw of them. He apparently got more satisfaction telling of the episode than looking for Clarkson.

2nd Contingent on Sunday afternoon parade, Karrakatta, 1900

In training at Karrakatta
Charles Mitchell second from left

On 7 November 1900 the 2nd Contingent, less a rearguard, left Cape Town in the transport *Woolloomooloo* and arrived on 8 December at Fremantle. They were disbanded on 7 March 1901. Mitchell was among volunteers to stay with the rearguard. The party was apparently attached to Lieutenant-Colonel Plumer, who had been given the job of cleaning up the isolated spots of resistance. The resourcefulness and natural horsemanship of the Western Australians would have appealed to Plumer. He was an outstandingly dashing young officer who had served with distinction in the Sudan and Matabele wars and had taken a leading part with the then Lieutenant-Colonel Baden-Powell in the relief of Mafeking.[9] The official records refer to an incident in these final cleaning up operations:

> At Klipplaat, on 5th February, 1901, Lieutenant Oliver had been sent in command of a small force of twelve of the 7th Dragoon Guards, twelve West Australians and three Cape Police to carry dispatches from Colonel Haig to another leader. The little party was surprised by several hundred of the enemy, they kept up a steady fire as occasion served from 11 a.m. until sundown, but eventually were all overpowered. Four Dragoons were killed and Lieutenant Oliver and several others wounded.[10]

Charles was attached to Lieutenant-Colonel Plumer as an outrider, and once, while carrying dispatches, was shot through the hat; he proudly displayed this garment on his return.

The war was over for Charles Mitchell. Of the Western Australians among whom he had fought, forty had been killed and eighty-six wounded. The services they had rendered were worthy of their home state and highly creditable to the individuals. They had won between them one Victoria Cross, two C.B., seven D.S.O., eight D.C.M. and three Mentions in Dispatches.

The remainder of the 2nd Contingent under the command of Lieutenant Duffy sailed for Australia in the S.S. *Tongarino* on 31 March 1901, on which day Charles was promoted to lance corporal. The troop ship arrived at Fremantle on Friday night, 19 April, and apparently anchored for coaling, either in Gage Roads or in the middle of the harbour.

The small Western Australian contingent landed by the steam tug *Awhina,* entrained at Fremantle and arrived at Perth Station at 1 p.m. They marched through the streets, behind a military band, and later were entertained at luncheon at Berry's Exchange Cafe, St George's Terrace.[11] The march drew a large crowd of sightseers, and youngsters found much to interest them in the appearance of a young Kaffir lad, walking alongside his 'boss' who had brought him back from South Africa. Complete with a wideawake khaki hat and haversack, he carried his carbine like a veteran. After the luncheon the troops marched to the drill hall, where each was advanced £10, then allowed to disband and depart with friends and relatives.

Lance Corporal Mitchell was discharged immediately on his return, receiving on discharge a final gratuity of £5, a clothing allowance of £1 5s 0d and the Queen's Medal with the campaign bars for Belfast, Diamond Hill, Johannesburg, Orange Free State and Cape Colony. The Western Australian Mounted Infantry had their own regimental badge—the initials W.A.M.I. cut into a Dutch two and a half shilling piece—one of which Mitchell retained all his life. His only grouch was that he had to pay his own train fare to Bunbury and arrange for his souvenirs to come down by a coastal trading schooner.

Souveniring was then either a fringe benefit to campaigning or a privilege peculiar to the rearguard. Stories vary as to what Charles brought back, from one to two light wagon loads, but it is fairly certain that there were a number of saddles, rhinoceros whips, Boer rifles and a Mauser pistol.

The rifles lay about at 'The Willows' until the family clean-up following sister Kate's death. To avoid trouble with the authorities for being in possession of unlicenced firearms, Nora (Mrs Rae) proposed they be buried in the sandhills. Her nephew, Guy Mitchell, came to the rescue; he knew a collector who would welcome the gift and with him they now repose.

Understandably, the war had a deep effect on Charles. He had been reared in an isolated community in which neighbourly co-operation was a by-word in a never-ceasing battle to harness a new and hard land. In South Africa he found himself pitted, with his comrades, against a dour, hard-working, patriotic people who, like his friends and family back at home, were concerned mainly with eking a living from the land. In the then admittedly infinitely stronger atmosphere of Empire solidarity they were enemies, but he was never really at ease fighting them. It was not his nature to pick on the underdog, and in later years he confided to a few intimates his regrets at having gone to the Boer War, not because he had not enjoyed the wonderful adventure and the manly excitement, but because during the fighting he had gained a great respect for the Boers and, even at that early age, had become convinced that they were in the right in the struggle.

'There should never have been a war', he said, on many an occasion.

1. His first appointment was with the Lands Department where he remained until 1894 when he was transferred to the Mines Department during the boom. For four and a half years he remained in the Mines Department and in 1898 he was transferred from the position of Under Secretary for Mines to the Aborigines Department as first Chief Protector. Henry Prinsep made a very positive contribution to our Public Service administration. Few people in the state could equal Henry's educational standard or cultural attainments.
2. Issue of Clothing and Accessories:

Jackets, drill, kh.	2	Spoon	1
Trou, drill, kh.	3	Holdall	1
Hats, F.S.	1	Knives, clasp	1
Caps, forage (glengarry)	1	Lanyard	1
Greatcoat	1	Kitbag, canvas	1
Boots, ankle, black	1	Razor and case	1
Guernsey, blue	1	Shirts, flannel	3
Blacking, tins	1	Socks, woollen	3
Braces, prs	1	Soap, pieces	1
Holder, button	1	Sponge	1
Brushes, sets	1	Towels	2
Comb	1		
Knife	1	*Weapons*	
Fork	1	M.L.E. rifle and bayonet	

3. Described in the *West Australian* (3 February 1900): 'Reserve of Officers, Captain in Command, Retired Officer of 21st Lancers, formerly 21st Hussars. Formerly Commandant of Western Australian forces'.
4. The 'Surrey Championship' of 100 yards, for members of contingents only, in uniform, was won by Western Australian, W. McKenna, prize money £2.
5. Dr David Ernest Williams was the District Medical Officer, Bunbury. He was born in 1868, the son of David Mark Williams, founder of the Liverpool Chest Hospital. Having qualified in Dublin in 1892, he practised surgery in Liverpool, then migrated to Western Australia in 1893, where he was employed by the colonial government.

   He was appointed Medical Officer, Carnarvon, and built the first hospital. In 1895 he transferred to Bunbury where he was instrumental in having the District Hospital approved, the foundation stone being laid in 1898. Subsequent to 1900 he was appointed Quarantine Officer in Fremantle.

6. Lieutenant J. (Jack) de Castilla, thirty-three years old, from Scotland, served two years as a non-commissioned officer in the 1st Aberdeen Engineer Volunteers and ten months in No. 1 Field Battery, Perth, as second lieutenant (extract from the List of Contingent in the *West Australian* 3 February 1900).

   Jack's family lived in Bunbury and were close friends of the Mitchells. Until he joined the W.A.M.I., Jack had never been on a horse. Charles taught him to ride, acting as tutor both at Karrakatta and in South Africa. Jack remained in South Africa, married, and lived there until his death.
7. This letter was reproduced in the *Bunbury Herald* in 1900.
8. Official records of the Australian Military Contingents to the war in South Africa, compiled and edited for the Department of Defence by Lt-Col. P. L. Murray.
9. Colonel Plumer's later career was equally as spectacular. After a term as Quartermaster-General at the War Office he served as commander of the 2nd Army group in France during World War I. He was promoted to Field Marshal in 1919 and created Viscount of Messines in 1929.
10. Official records of the Australian Military Contingents to the war in South Africa.
11. Situated on the ground floor of Emmanuel Bros Buildings, on the corner of Howard Street, now occupied by the Perpetual Executors, Trustee and Agency Co. (W.A.).

7

Post-War Years and Marriage

The aftermath of war was the same for Mitchell as for any returned serviceman. Having regaled his family and friends with his experience, having spent all his money, and having been restored to full bloom with his mother's cooking and sisters' care, he found himself face to face with the necessity of finding a job. As on a previous occasion, he found a berth with one of his brothers.

Some time after leaving school his elder brother James had joined the Western Australian Bank as a junior in the Bunbury branch. From there he was transferred to Geraldton for a number of years before being given charge of the Northam branch, which position he occupied until his election to the Legislative Assembly as member for Northam on 27 October 1905.

While at Northam James had taken a keen interest in agriculture, acquiring suitable lands and supervising their development and cultivation. He was accounted one of the most practical and energetic agriculturists in the Avon Valley. He farmed on a large scale and in 1906 had 1,200 acres in crop.

Charles elected to go and work for James at 'eight bob a day and tucker yourself', as opposed to the offer of a neighbour—Mrs Maud Dempster of 'Buckland'—of 'six bob a day and found'. James, at that time, had a number of properties, either owned or under lease: one, 3 miles east of Northam, between the salt lake and the Kalgoorlie railway; a second, part of the Seabrook Estate;[1] and the third on the main road, 3 miles west of Grass Valley.[2]

Oaten hay from the Northam district was of prime quality and in strong demand. Chaff cutting with horse works was one of the principal spare time items of employment and, under James's supervision, Charles worked hard. While responsible also for overseeing all employees, he never achieved full managerial status—that was to remain this particular proprietor's constant prerogative, whether as farmer, banker or parliamentarian.

By application and knowledge of farming James had made a reputation for himself as a country banker, in both the obtaining and control of business. He was displaying early signs of becoming a law unto himself. He disliked music, and when the piano was played, whether company was being entertained or not, he would leave the room. He participated in no sport, yet enjoyed watching. He was dogmatic, and when he made up his mind to do something, he did it. An instance of this was his preferment of the young officer, Harry Chatterton Nankervis, who eventually married the Mitchell sister, Marion.[3]

There was one particular enjoyment now available to Charles; he found time in the weekends to play football, and became a playing member of the Federals Club, one of the two Northam teams.[4] He played in defence and as a follower, and while seldom mentioned for brilliant or even outstanding play, was invariably among those named for 'hard work' or 'giving useful service'.

His style was marked with zest and determination, which earned him the nickname 'Fireworks'. To match his strength and forceful exuberance, an opposing team recruited a professional pugilist who, as the game progressed, made no effort to conceal the fact he was there to impede Mitchell. After repeated clashes and a few sly blows, Mitch enquired:

'Are you looking for a fight?'

'Yes', was the reply. 'If you think you're good enough.'

'Right!' said Mitchell, 'I'll take you on after the match, behind the dressing shed.'

Word of the arranged bout got around and soon someone whispered in Mitchell's ear how he had been set up. He only nodded. When the match was over and the crowd had gathered to watch the fun, he walked up with two pickets under his arm. He handed one to the 'pug' with the explanation: 'When I said I'd fight you, I didn't say what with. Now we're evenly matched.'

'Get to hell out of it!' said the startled pug. 'They told me you were mad, but not *how* mad!' He stalked off with his disgusted, back-patting supporters, and the Federals' honour was satisfied.

James had married Clara Spencer of Bunbury in 1888, and when Charles arrived in Northam, there were four children: Robert Francis Purnel, Roy Bedford, Gerald James and Jean Clara.[5] Clara and James entertained a great deal and so to assist with the children and the maintenance of the large residence,[6] they employed a lady help, Constance Rossiter, who had come to

Western Australia from Victoria with her sisters Maud and Ida Marian, in search of employment.[7]

Ida had been planning to become an Anglican missionary, but became instead her sister's employer's sister-in-law. She had come to stay with Constance to recuperate from a bout of typhoid fever, for which part of the treatment was to shave the patient's head. Charles always averred that it was 'Mother's mass of tight fair curls, all over her head, just like a baby's' that caught his eye and heart. His family thought that her serene smile and placid nature made quite a contribution. It is reported that during the winter months of 1904, while the young lovers would be holding hands on the sofa in front of the living-room fire, James would come in—reputedly never later than 9 p.m.—and with the apology: 'You two don't need this any longer', would remove the wood and coals to his own bedroom on a large shovel. One wonders whether he opposed the possible union, or merely wished to ensure that one of his employees would retire early in preparation for the following day's work.

Charles and Ida were married by the Reverend (later Dean) Robert H. Moore in St John's Church of England, Northam, on 13 June 1905. Charles was twenty-six, Ida, thirty. The reception was held at the home of Mr and Mrs David Morrell,[8] who had become great friends of the Rossiter sisters. James, anxious to retain his brother's services, had a three-roomed, mud-brick house built on Location 2 of the Grass Valley property,[9] and following the honeymoon, Charles moved there with his bride.

There and then began a wonderfully happy life for Charles. Ida adored him, understood him and spoilt him utterly, and Charles loved and cherished Ida no less in return. Consequently he always spoke highly of the state of matrimony, advising all young fellows in its favour, averring they should be ashamed of themselves for being single, and that they should not wait till they thought they could afford it.

'When I was married, all I had in my pockets was two bob and a 4oz plug of tobacco, and look at me now', he used to say, and then he would add: 'Don't be afraid of big families; they look after themselves. The elder takes care of the one just below and so on. Great training!'

THE HONEYMOON BROUGHT THE NEWLY WEDS HOME to Bunbury. The first weekend they were home was the occasion of a meeting of the local Beagle Club, of which William Bedford was one of the main patrons and an active supporter.

Everything was carried out in the best English tradition; there was a Master of the Hunt and a pack of beagle hounds. All that was missing were the red coats and a fox. A brush wallaby, of which there were plenty, was the accepted substitute. The hounds got the scent, gave tongue and the hunt moved off—not at the break-neck speed of its English counterpart however, but at a sedate canter. The restrained pace galled Charles, the youthful former mounted infantryman, not permitted by protocol to pass the Master. He caught the eye of equally young H. R. (Harry) Rose of Burekup,[10] who nodded in understanding. They checked their mounts until the hunt was out of earshot then at full gallop swung far round to the front, caught the brush and when the hunt eventually arrived in all due decorum, there was the 'fox' tied to a tree, torn to pieces by the hounds, which stood around literally licking their chops.

William Bedford was livid. Oh! the shame, to have reared a son with so little respect for tradition and for his elders. Why, he would be shooting a sitting duck next!

Charles and his bride left Bunbury the following morning, by invitation; and it is assumed, from the reports of his financial standing, that the honeymoon also ended there and then. He did not again see his father alive. William Bedford died at 2 a.m. on 1 April 1907, aged seventy-four years. He had been in feeble health for some time, not having recovered from the shock he had sustained in falling into the Brunswick River some months before.[11]

His wife was the sole beneficiary and his eldest son, James, was named executor. Caroline considered she had been more than well provided for, and 'Kelly's'—the farming property of some 400 acres acquired in 1901 adjoining the town of Donnybrook[12] —should have been left to sons Harry and Charles. She said that if they would farm it in partnership, the property would be theirs in her will.

Harry had built a large brick home some years before on Wellington Location 455, adjoining Kelly's, which stands to this day and is still the family home, occupied by the widow of Harry's eldest son, Jim. She was Mollie Pell, of an old Goldfields family. There were four children to their marriage.

Caroline's offer was of great interest to Charles. Gold had been discovered in the Donnybrook district in October 1899 by R. L.

Hunter, his son R. L. Hunter, Jr, and Robert Sykes. It is believed that the original discovery was made on private property some $2\frac{1}{2}$ miles west of the town.

The Donnybrook goldfield was proclaimed on 17 November 1899, to take effect ten days later. W. A. G. Walter, resident magistrate at Bridgetown, was appointed warden (without salary), with A. L. G. Lefroy as temporary mining registrar.

In spite of several locally organized representations, the last recorded in 1902, the government declined to offer the customary discovery reward, being of the opinion that the result of the field did not warrant such action, although up to 200 men worked on it at one time. Evidence of old workings still exist, south of the town, and occasional tourists seek them out. The goldfield was cancelled by notice in the *Government Gazette,* 6 March 1908.[13]

However, Donnybrook had boomed by reason of the gold find, with the Railway Hotel proving a profitable venture. There was also a demand for building blocks and William Bedford, aided by his son Harry, had subdivided the southern halves of Wellington Locations 90 and 242 into town blocks.[14]

Mother's word was good enough for Charles, and the promise of a secured future was in fact timely and heartily welcomed. Ida had lost her first child, a son, soon after birth, and a change of scene was desirable. They left Northam and James's employ practically immediately, and took up residence in the cottage at Kelly's. By mutual agreement Harry would continue with his profitable contract surveying venture, while his family lived on the farm.

When Charles left for Donnybrook he was followed by two of his men, Harvey and Hindes, who were great workers. They were to receive the same pay, and they were fond of the 'young boss'. Hindes was a ploughman, and Donnybrook people relate to this day his record-breaking exploit when, with a single furrow No. 15 Massey plough and two horses, he ploughed $2\frac{3}{4}$ acres in one day. Harry measured and certified the area.

Running stock, and buying and trading to supply the local butcher were the principal functions of the enterprise. Charles did the buying, selling and slaughtering, and in his spare time, managed to re-fence the whole of the property.

Harry's eldest boy, James Henry Bremer, who was born in 1902 and who died recently, remembered vividly one particular demonstration of the prowess of his hero, Uncle Charles. The property being close to the town, the sheep were subject to considerable molestation by the town dogs. At first light one summer morning Mrs Harry Mitchell, from the eminence of her kitchen window, observed four of them worrying the sheep. She immediately beat the iron triangle, normally intended for calling the hands to meals. Charles 'got the message' and, as Jim recounted:

> Then up from the cottage, which was situated on Wellington 450, to the left of the bridge at the end of today's Kelly Street, came Uncle Charles, mounted bare back on his favourite bay gelding Nemo. He circled between Harry's house and the river at full gallop, and with four shots from a Mauser pistol [a souvenir of the South African War], killed three of the dogs and seriously wounded the other.[15]

Charles still enjoyed the thrill and clash of football, and in his new location played for Bunbury. His contemporaries described him as strong and untiring. 'A good man to have in the side; he always made it hard for his immediate opponent'—was the general verdict. Selection to play for Bunbury meant a long day, leaving home at 7 a.m. and returning about midnight. He loved the game, but at the second match of the season he arrived home disgusted. 'There was no match', he complained, 'Busselton didn't turn up.'

He never bothered to go in again and concentrated on his second love—polo—of which he had become a devotee during his service in South Africa. Nemo was a splendid pony, capable of pulling a sulky anything up to 30 miles to a match, then of carrying Mitchell, a vigorous and fearless player, throughout the whole of the game.

On 31 July 1907, to their great and mutual joy, Ida presented Charles with a redheaded baby girl, Constance Joan. Up till then life had not been complete for the young couple.

In spite of Charles's hard work and, it is believed, solid trading, it was becoming increasingly apparent that there was not sufficient return from the farming venture to maintain two families. Charles saw the answer to his problem in an advertisement inviting applications for the position of inspector in the Agricultural Bank. He promptly applied for the position and shortly afterwards received a vist from William Paterson, the managing trustee of the Agricultural Bank of Western Australia. Paterson, recognizing in Mitchell the latent talent which the Bank could use to its advantage, there and then offered him the position of inspector, to be stationed at Narrogin, on a salary of £200 per year, with a range of up to £240, plus a fodder allowance for his horses. Charles talked the offer over with his wife and Harry. He accepted it and moved to Narrogin, commencing work on 1 August 1908, the first full-time field inspector of the Agricultural Bank.

At the age of twenty-nine he entered on his real life's work.

1. Lots 4, 5, 6, 7, 8 and 9 of Avon Location K, 1,345 acres.
2. Portions of Lot 2 of Location N.1 and Location 61 (each of 226 acres). Two adjoining lots, Throssell Estate 79 and 80, of 134 acres, were acquired in 1911.
3. A young officer, Harry Chatterton Nankervis, had been transferred to the Northam branch from Perth and became the equivalent of today's manager's clerk. James took a decided fancy to 'Nanky', asking him not only into his home, but without authority paying him the 'goldfields allowance' of £50 per year, stating that a young man could not live adequately on a salary of £120. Nanky was an excellent lacrosse player and James, again on his own initiative, used to permit his taking those Saturday mornings off to travel to Perth when the Bank's team had a particularly hard match coming up.

   There was consternation when the Bank's inspector visited Northam and became aware of the extra payment being made to the manager's clerk. Not even the state manager was prepared to go against James's decision, so it was legalized by the transfer of Nanky to the Goldfields.

   After some years in various goldfield towns he returned to Perth in 1907 in time to play the season with the Perth Club, W.A.N.F.L., and assist them in winning their first and only premiership till 1955. Nanky played as half-back and ruckman. The following year, by design or accident, he was transferred to Bunbury, renewing acquaintance immediately with the Mitchell girls whom he had met previously in Northam.

   He captained the South Bunbury team, in which Tom Mitchell played, to a premiership and found time to court and win the hand of Marion (Win) who, when he knew her in Northam 'cheekily turned her head in the air whenever she passed Nanky'. Win confessed in later years it was because she then considered him spoilt. They were married in Bunbury by Bishop Goldsmith, first Anglican Bishop of Bunbury, on 6 July 1909. Marion drove to and from the church in Sir Newton Moore's phaeton, driven by Sir Newton's Aboriginal coachman 'Pidgeon'. The reception was held at the Willows.
4. The district teams of York and Newcastle (now known as Toodyay) made up the balance of the Association.

5. Married Terence Oliver Anketell on 15 August 1928 at St Patrick's Church, Perth. A son, Richard James, was born on 1 June 1932.
6. Two-storeyed brick dwelling, on the corner of Hawes and Duke Streets. Owned today by Harley Colebatch and rented as flats. Hawes Street has a unique distinction in that two of Western Australia's Premiers have resided there: Sir James Mitchell, and the Hon. A. R. G. Hawke, Labour Premier (1953-59), at No. 16.
7. The Rossiter sisters were born in the village of Cretchel, Dorset, England, to Edward Rossiter and Marian Eliza Hodges. Following the death of his wife, the father migrated to Victoria with his three daughters and later re-married.

   Constance Ayles Rossiter later married Sidney Frederick Morrell of 'Fair View' near York. She died giving birth to their first child, James Rossiter, who today (1969) is the Methodist minister at Wongan Hills.

   Maud Mary Rossiter, a school teacher, married Lewis George Wilcher, superintendent of the Swan Boys' Orphanage, on 21 December 1907. One child, Lewis Charles, was born of the marriage on 9 December 1908. Subsequent to the death of her husband on 30 November 1909 Maud went to live in South Australia. Apparently having qualified as a trained nurse whilst at the orphanage, Maud obtained employment in the then Children's Welfare Department—now known as the Department of Social Welfare. On 1 August 1916 Mrs Wilcher transferred from the Welfare Department to the South Australian Police Department. She was the third woman police constable appointed in that state. When appointed principal of the women police on 1 August 1935 she held the rank of first grade sergeant. She resigned from the Department on 30 June 1940 at the age of sixty.

   Her son Lewis, after a brilliant scholastic and athletic career at St Paul's Day School, Pulteney Grammar School, St Peter's College and the University of Adelaide, graduated B.A. with first class honours in History and was selected as S.A. Rhodes Scholar for 1930. Entered Balliol College, Oxford, and earned M.A., B.Litt. After a varied and colourful academic career, he was honoured with C.B.E., 1955; and he was appointed Warden, Queen Elizabeth House, Oxford, 1956.
8. Parents of the late Tom Cloudesley Morrell, born 18 September 1893, who was for many years manager of the Commonwealth Bank in Narrogin and life-long friend of the Mitchell family. He died 6 July 1967.
9. The walls of the house, which was known as 'The Wattles', still stand. The majority of the mud bricks, reinforced with straw, remain quite sound. The interior walls, first covered with a lime wash and then trowelled, presented a hard smooth finish, still visible. The

property is today owned by John Henry Clarke, who acquired it in 1955 from Allen Joseph Sermon.

10. Rose was later to be chairman of the Bunbury Race Club and was a member of the Group Settlement Advisory Board.
11. The obituary notice mentioned, *inter alia*: 'He was for some years handicapper for the Perth Racing Club and also acted in a similar capacity for the Bunbury Race Club. Mr Mitchell was also credited with being familiar with the pedigrees of all the old country racing stock for many years.'
12. Comprising Wellington Locations 90, 242, 450, 4057 and 4058.
13. Production figures were: alluvial 67.25 fine ozs; dollied 43.08 fine ozs; ore 1732.80 tons for 831.94 fine ozs; total 942.27 fine ozs. These figures are a grand total; there was production as late as 1936.
14. Those sold from Location 90 were built on and are still occupied as part of the townsite; but with the short life of the goldfield, little building was done on the subdivision, and over the succeeding years Harry's son Jim had repurchased the blocks, including them in the farm.
15. Personal communication.

8

His Work 1908-1921

The following four chapters deal with Mitchell's life and work in Narrogin from 1908 to 1929. For the purpose of clarity they have been arranged to keep his private life separated from his official undertakings. Chapters 8 and 10 deal with official matters, and 9 and 11 show Mitchell as the family man, the neighbour, and the citizen.

Mitchell shared the crossroads in his life with the state he was to serve so well. In 1908 declining gold production had resulted in unemployment on the goldfields and throughout the state generally, and the logical alternative was to extend primary production.

In Charles's elder brother James the right man was available to formulate the fresh and vigorous policy so necessary to meet the challenge of the changing times. James Mitchell had entered parliament in 1905, aged thirty-nine, as a National Party[1] candidate. Backed by his energy, zeal and knowledge of agriculture, he was appointed Minister-without-Portfolio in 1906 to assist the Hon. Frank Wilson, Minister of Agriculture in the government headed by Newton Moore. The Moore government was returned in 1908 for the second time, with young Mitchell clamouring for recognition and the chance to solve the state's most pressing problem. The two years he had spent in close association with the administration of the Department of Agriculture had not been wasted—it had enabled him to obtain additional technical information and the time to perfect his plans for increasing wheat production.

Newton Moore resigned as Minister for Lands on 30 June 1909 and simultaneously the Hon. Frank Wilson relinquished the Agricultural portfolio. On the same day James Mitchell took over the two portfolios. He had already decided upon the type of organization he intended to set up, and as he knew what he wanted to do, he saw no need to appoint advisory committees.

He was a strong man who brooked no interference. A contemporary wrote: 'James Mitchell then brought together idle land and idle people.'

An early admirer of Sir John Forrest and later a close friend, James had been intrigued by and subsequently had studied the 1892 political aspirations of Sir John, then Premier, to settle the colony with a 'bold Yeomanry'. Now came the opportunity to advance those earlier ideals, remoulded in the light of experience gained, plus some of James's own ideas.

How well he succeeded is remembered by his sister Nora, who recalls his elation when, in 1912, Forrest (by then Lord Forrest) complimented him: 'James, please accept my congratulations. You've accomplished greater success with the settlement of our land than I ever managed.' It was in this era that James coined his first recorded epigram, to emphasize the shift in Western Australia's population from the goldfields to the farming areas: 'Gold brought them to Western Australia, superphosphate kept them here.'

Before 1908 a line on the map drawn north and south through Meckering represented the accepted eastern boundary of the wheatbelt. This line of demarcation most probably was prompted by the caution of financial institutions chary of lending for farm development east of it. If one wanted to settle in the yet unproven eastern lands it was at one's own risk, the Lands Act permitting 'selection before survey'. Under this form of selection a would-be farmer or grazier would normally ride through the bush lands until he found permanent water. He would select sufficient land to encompass and control the natural water supply, make application for the land by description, and await the arrival of a surveyor to fix the boundaries. Such permitted action, costly and time-consuming to the state, gave to the 'selector' a pre-emptive right to grazing the adjoining country, with or without an official lease, and eventually to pick the eyes out of the nearby farming land for development as inclination and financial reserves permitted.

The Coolgardie Water Scheme,[2] completed in January 1903, gave an assurance of domestic supplies along its route, and enterprising farmers—ignoring general predictions of failure—pushed wheat growing east of Meckering.

Messrs Packham and Donnan at Tammin, Leake at Kellerberrin, Growden at Nangeenan, and Forrester and Nunn at Southern Cross all reported having grown profitable wheat crops, and in 1904 the Lands Department commenced clearing operations on an experimental farm at Nangeenan as a model for settlers taking up nearby land. It would probably have been discontinued as soon as the neighbouring farmers were on their feet, but the Agricultural Department took it over in 1907 as an Agricultural Research Station. The first crop was grown in 1906, and the farm manager reported that 156 acres had been cut for hay, yielding an average of 11 cwts per acre. 'For grain a new wheat hybridised by Mr. Berthoud of the State Farm, Hamel, yield 14 bushels, and "Yandilla King" 18 bushels', his report stated. 'All plots received 75 lbs of superphosphate per acre.'[3]

James Mitchell's first action was to extend the eastern boundaries of the wheatbelt to east of Burracoppin, where the limit of extension was automatically fixed by the No. 1 Rabbit Proof Fence, begun in 1901 and completed in 1907. The Land Act, amended, no longer permitted 'selection before survey'.

All available contract surveyors were engaged, and each was assigned an assistant capable of handling a theodolite. Before long there were seventy-four theodolites in the field, mapping out land. From 1 July 1908 to 30 June 1914, 12.375 million acres of agricultural land were surveyed, representing 24,399 holdings or reserves.[4] The fixing of boundaries for 5,168 holdings, totalling 2.6 million acres, represented the first year's effort under Mitchell's regime as Minister for Lands. The greater part of the holdings newly created over the six years were 1,000-acre blocks in the eastern wheatbelt, with the soils thereon classified as first, second and third class, and the relative areas recorded for each block. The impetus given wheat farming from this seemingly unlimited availability of land—in which Mitchell bore his share—resulted in the trebling of Western Australia's wheat crops between 1908 and 1912, an increase of half a million acres.[5]

It was in this challenging era that Charles, aged twenty-nine years, emerged as a bank inspector, developing the rule that every man was entitled to a 'fair go'. It was to be an era of unequalled change and progress, but he little knew then that he would meet it and instinctively adapt to it, mould the new administrative techniques, and later, by precept and example, extend his philosophies to the greater part of the wheatbelt.

The Narrogin district of which he took charge extended from a north-south line through Collie in the west, and eastward as far as settlement existed. The northern and southern boundaries were Pingelly and Lime Lake. His appointment was timely and urgent. Special settlement schemes at Yealering and Dorakin had been declared in 1908, while allocations of Crown land for Narrogin district in February of that year totalled 40,132 acres for 100 applicants. Government policy was to push settlement through the opening of new railway lines. The Collie-Narrogin and Wagin-Dumbleyung railway lines had been opened in 1907. The Narrogin-Wickepin line, then in course of construction, would be completed in 1909.

Bruce W. Leake, in his *Reminiscences,* makes an interesting observation of this particular aspect of expansion:

> There was considerable argument as to whether the country could afford to lay the necessary railway lines and it was not until Smith and Timms, railway contractors, stated in newspapers that they would build those for one thousand pounds per mile, that the Government made a start.[6]

The Agricultural Bank Act proclaimed in 1894 led the *West Australian* to state: 'Undoubtedly the Bank stands unique in colonial representation.' It was unique in that its function was to advance moneys to persons who had no other security than the block of land they had selected and were prepared to develop and settle on.

The government had accepted that it was not possible for the average man to develop farming land solely from his own financial resources. The initial capital made available to the Agricultural Bank was £100,000. Total advances per farmer were limited to £400—against 50 per cent of the value of clearing, cultivating or ringbarking.

In 1896 the maximum advance to a borrower was doubled and the Bank's power extended to lend up to two-thirds of the value of clearing, cultivating or ringbarking, plus the added items of fencing, buildings, water supply or draining.

This practical official approach to land settlement continued to take shape, and the right of overcrowded people to settle in undeveloped lands beyond their own boundaries was given effect by the passing of the Agricultural Lands Purchase Act of 1896. This provided for the re-purchase of estates, invariably those in whole or in part subject to the original free land grants of the colony, which had not been developed sufficiently to warrant their retention in the hands of the original owners.

Demand for assistance was such that within three years the Bank's capital had increased to £200,000, and in 1902 the loan limit was raised to £1,000, with added power to the Bank to lend for horticultural and viticultural pursuits, and to discharge farmers' outside liabilities.

William Paterson, the member for Murray in the Legislative Assembly from 1890, had been a strong supporter of the Premier, Sir John Forrest, in his efforts to establish the Agricultural Bank. In this he was in close association also with the members for Fremantle and Northam, W. E. Marmion and George Throssell. He retired in December 1894 to take up the management of the Bank, which he held until his death on 11 March 1920. A man without banking experience, he nevertheless was an extremely practical agriculturist, and up to the end of 1906 travelled extensively, interviewing applicants for loans, inspecting work done, assessing progress, providing advice, and instilling confidence.

It is said that some of Paterson's trips by horse and sulky lasted up to two months. He was a familiar and welcome visitor in all parts of the new settlement then so rapidly spreading out over the Western Australian countryside; it was his bank, and he had to know what was needed to make it tick. There is no record of his having had any assistance except of a part-time nature, accorded by Lands Department agents and inspectors stationed at vantage country centres.

When the Bank became a body corporate in 1906 William Paterson became the managing trustee, and received assistance from part-time trustees, the Hon. A. R. Richardson, I.S.O., and Frank V. Cooke. At the same time the Bank's capital increased to £1,000,000. With the incorporation of the Bank, it became most patent that Paterson, as managing trustee, should now concentrate his skill and experience on the management of the rapidly growing institution. To permit of this, A. G. Hewby, Inspector of Lands, Lands Department, was appointed chief inspector, Agricultural Bank, on 30 April 1907.

Hewby's investigations and observations over the next twelve months confirmed Paterson's views that it was high time the Bank acquired its own personal full-time field staff. They both knew the type they wanted: men with whom the farmers could identify themselves, men who spoke the farmers' language. Mitchell was the first appointee, ten other field inspectors being added to the chief inspector's staff by 1912.[7]

In 1909, by which time Head Office staff of the Agricultural Bank totalled twenty-four (twenty-one males and three typistes), approval was given for advances to be made for purchase of farming machinery manufactured in Western Australia, reflecting support of the government-owned Implement Works.

Statistics are not available as to the proportion of farmers who sought and received Agricultural Bank assistance, but it was accepted as general practice, and one not discouraged by the private banks, which maintained the credit account only—with little risk and less responsibility.

A new settler did not have to wait for the arrival of an inspector to lodge an application for assistance, however. There were no needless delays; the farmer either made contact with the Lands Department agency or Lands Inspector, or wrote to Head Office, which had already received particulars of his holding and its soil classification from the Lands Department. Even in Mitchell's district loans were approved and documentation completed, by post, without prior reference to the inspector.

Further streamlining provided that a certificate from a neighbour was sufficient basis for payment to be made against improvements completed, and for stock and plant already delivered. An inspector received notice of all such payments and in due course arrived to check their accuracy. One could conclude that this 'honour system' must of its nature have had some shortcomings; but, all-in-all, it worked very well.

One story is told of a resident from the Bruce Rock district who called on the general manager, William Paterson, to report that 'so and so' had been paid for 135 acres of clearing he had never done.

'Impossible under our procedure', snapped Paterson. 'Wait till I get the file and I'll show you.'

The file arrived, disclosing that payment had been made under a neighbour's certificate. Paterson could hardly believe his eyes.

'Didn't you say your name was B——?'

'Yes.'

'J. W. B.——?'

'Yes.'

'Well, you're the one who signed the certificate!'

'That's correct, that's how I know.'

In later years Mitchell related a story of another alleged overpayment on clearing which Head Office was striving to recover. The farmer refused to be coerced, claiming that there was no error and, in any case, he proposed doing more clearing when he could find the time. When at last the final letter was sent:

'Pay up or else'—the exasperated farmer replied: 'I have your letter of the ——th before me. In a few minutes I'll have it behind me. Yours faithfully.' No record is held as to whether the impasse was ever resolved.

To service the demand for the land, the Lands Department had opened a Branch Office in the Agricultural Hall in 1908. Mitchell's first assignment was to settle successful allottees on the Dorakin Estate. It was two weeks before his family saw him again.

BANK INSPECTOR MITCHELL TRAVELLED by buggy and pair, harnessed in tandem. The horses were named Nemo and Nominus by his wife, an ardent reader of the classics. Picture Mitchell leaving home for a trip in a 'King of the Road' buggy—a high-wheeled American type introduced originally for the goldfields, but from which Mitchell had removed the steps to prevent entanglement with the scrub—loaded with a bag of chaff, a riding saddle, camping equipment, food and papers. When he arrived at his destination one horse was saddled, the other was hobbled, and the day's work began.

Mitchell was ever practical and conscious of economy. During his final week away, as the load of chaff lessened, he would begin to cut stove-length firewood, stowing it to full capacity, for his return.

By arrangement through Head Office, and on request, the Railway Department would provide a van into which Mitchell could load his buggy, horses and impedimenta between sidings, the van being shunted on to the spur line till ready to be picked up. He did not favour this method of travel, however. Seldom did the railway schedule coincide with his need to move elsewhere, and delay irked him. He preferred to spend the time driving the long distances rather than to sit about and wait.

From the outset Inspector Mitchell fixed rigorous and unswerving principles. At no time would he accept meals or accommodation from an Agricultural Bank client, a restriction which applied equally to his field staff. He could neither bring himself to impose on them nor to be placed under an obligation. Irrespective of the weather, he made camp and prepared his own meals, courteously and diplomatically adhering to his principles. Mrs Hugh Henderson[8] of Dudinin recalls his declining to dine at her girlhood homestead, saying, 'the women have enough to do without having someone extra'. This rule could, however, be relaxed in the case of bachelor settlers, Mitchell being quite happy to share their camp fires and his or their tucker, but seldom the shelter of their tents or humpies. Company was always welcome to the lonely settler—Mitchell's more so. An accurate shot with the small bore rifle he carried, he always had fresh meat—parrots, brush 'roo or rabbit—to contribute, which made a welcome break from the usual diet of 'bully' beef.

He was a natural bushman: never lost, never at a loss, and capable of living off the country when need dictated. Camp consisted of making a hollow for his hip alongside the camp fire, a ground sheet and blankets, with a tarpaulin to pull over him if it rained. Except for a period between 1916 and 1920, when he rigged a bunk across the seats of his Model T Ford, he made no concession to his material comfort when camping. This was no pose—it was what he had grown up with and liked. He could never concede that he was not capable of roughing it as he had in his youth, but the explanation he gave was a boast rather than a reason. 'You can get up at daylight and get on with the job, instead of hanging round half the day waiting for hotel breakfasts.'

One young, newly appointed inspector, who was on his first trip with C.D.M. (Charles's reference among his adoring staff), never forgot the scorn Mitchell displayed when he changed into pyjamas, then the look of absolute disgust which came over the boss's face before he pulled the blanket over his head to hide the sight of one of his field officers putting sheets on his camp stretcher. This point is stressed, as Mitchell normally never pulled the blankets over his head, even when it rained; moreover, he invariably wore his hat to bed.

LATE IN 1908, WEST OF THE 101-MILE GATE (101 miles from the southern coast on the Rabbit Proof Fence) Mitchell met a young boy in long black stockings trudging along with his worldly goods tied in a red bandana handkerchief and carried over his shoulder at the end of a stick. Mitchell stopped.

'Son, do you want a lift?'

'Yes, please, sir.'

There and then began a life-long friendship. The boy was C. G. (Clarrie) Russell, aged eleven, who, with his mother's blessing, had left his stepfather's farm at Noman's Lake to seek his eldest brother Arthur, who was reported to be working for Harry Mullan, father of E. J. (Jack) Mullan of Dudinin.

The boy had walked to Mullan's on the first day, but learned that Arthur was sinking a government dam at 'The Paper Barks',[9] 20 miles further east. Having stayed the night with Mullan, Clarrie had walked 11 miles next day before he met up with Mitchell, who took him on to the Paper Barks where the boy was engaged at 12s 6d per week, as cook and spare-time driver for the dam-sinking teams.

Boys were men in those days: Clarrie relates that his other brother, Wally, aged fourteen, was cutting jam posts at that time with Ted Plant of Noman's Lake, and considered it a poor day's work in which they could not each cut a hundred posts.

Mitchell came to take an increasing interest in his protégé and never failed to call and see him when in the area.

As soon as Clarrie was old enough to acquire land he took up a block at Geetabin, north of Jitarning, and applied for Agricultural Bank assistance. Clarrie related that his two-roomed hessian hut was one of the few places at which Mitchell ever stayed overnight, and then only on the occasion when he called in the company of E. B. (Bertie) Johnston, M.L.A. for the district.[10] After staying the night Clarrie, Mitchell and Johnston went 'rooing next day.

He further related Mitchell's standard procedure of always inspecting and measuring improvements for which loan moneys had been provided, before contacting the farmer. This was the yard-stick for assessment of effort and application. The completed works, good or indifferent though they may be, enhanced or leavened first impressions or personality clashes.

His fieldwork was never skimped; guesswork or 'near enough' never applied. Measurement was by pacing, and Mitchell had the exceptional gift of being able to maintain the same length of pace, an exact yard—a rare feat—not only over periods extending up to twelve hours a day, but also over the period of thirty years. Few ever kept up to him in a day's walking, or tried. Also, he had a photographic memory of any property he had visited: there are recorded examples of his quoting accurately the chainage of certain fences and areas of paddocks although he had not been on the properties for over thirty years.

The usual practice for an inspector, on receiving appointment, was to lay out, by tape or surveyor's chain, an exact ten chains, marking each end with a peg. Then by walking backwards and forwards between the pegs, he could determine on average how many steps it took, pacing naturally. It was amazing the accuracy one developed with practice. My own stepping was 225 paces to the ten chain.

No farmer, on first contact with Mitchell, would guess his thoughts. His somewhat abrupt manner and terse speech—to which could be added a gruffness, designed to hide his innate kindness, softness and interest in others—gave nothing away. Few ever knew him till they were in real trouble and needed help as never before; and if they were worthy of it, they got it.

One settler, then living in a tent to conform to the Lands Department's residential requirements, and endeavouring to develop his block between seasonal work on his employer's nearby farm, now relates the first and subsequent meetings which governed his farming life and eventual success.

'I've had good reports about you', said Mitchell, on introducing himself. 'You've got a small block but I'll give you a go.'

It took this settler four years to get ready for his first crop, but how was he to finance it? Mitchell, whom he had not seen for more than two years, suddenly appeared.

'You'd better apply for I.A.B.'[11]

'I'm not eligible', was the reply.

'Never mind', said Mitchell, 'You need it. Apply, and while you're on the job, put in for more land. I'll back you up.'

Some seven years later, concerned with his mounting debt and inability to meet his obligations, the settler travelled to Narrogin to place his problem before Mitchell.

'Stop worrying, that's my job', Mitchell said. 'You haven't got a farm as yet. Keep on going the way you are.'

It is possible that the abnormally bad season in 1911, coinciding with a change of government,[12] actuated the shift in political thought which prompted the proclamation of amendments to the Agricultural Bank Act in the following year.

The amendments increased to £2,000 the maximum advance available to each borrower. The percentage paid against the value of improvements was raised from two-thirds to a full 100 per cent, and no restriction was placed on the manner in which the amounts were to be advanced.

The prior establishment and maintenance of an equity by the farmer borrower cannot, in light of experience, be overstressed.

Protection of his own as well as the government's contribution promoted a different outlook in the farmer.

The effect of the 1912 amendments was to convert the Agricultural Bank from an 'improvement' to a mortgage bank.

The Royal Commission's report of 1933 emphasized that no losses of any consequence had occurred in the face of the Agricultural Bank's operations from 1894 to 1912 nor were any envisaged. In point of fact the first losses recorded were £90 6s 0d in 1911 and £143 11s 6d in 1912. With the exception of an amount of £7 10s 0d interest written off in its early stages, these amounts represent the first trading losses which the Bank had to record since its inception.

The increase from £1,000 to £2,000 maximum advance in itself was warranted. The trustees report for the year ending 30 June 1911 recorded that the Bank had opened 1,271 new accounts during the year and closed 1,050, mostly borrowers who had transferred their securities to private institutions. The report continues:

> Our experience is that as soon as the full value limit is approached, borrowers seek accommodation elsewhere. The banking facilities which these Institutions can offer, and an easier access to credit, particularly in respect of short loans for immediate necessities, outweigh the advantages of a fixed mortgage though carrying a lower rate of interest.

With the coincidental waiver by the Lands Department of the stipulation that the farmer be fully resident, the state witnessed between the years 1912 and 1929 the transition from pioneering to speculative farming. With no equity required, many farming-minded persons, having nothing to lose but a few years' work—and much to gain—took up land and hoped for the best. Bad times or bad seasons could bring about a near general exodus.

This liberalization of 'farm credit' typified the growth of state government assisted land settlement throughout Australia, which probably, at this point of time, reached its peak in Western Australia, where a maximum of available agricultural land called for the maximum effort in development from the government in furtherance of its aims. These aims were the extension of the wheatbelt and the increasing of the rural population, thereby contributing toward first, the state's self-sufficiency in wheat and coarse grains and second, to exportable surpluses. It also called for more effort from the man-on-the-land in return for the government's initial assistance with liberal, long-term finance.

To sum the situation up: the Land Board selected the settler and the Agricultural Bank had to finance him. The settlers were recruited entirely from within the state—few were trained farmers, many had neither experience nor private resources, and some were quite unfitted for the life. Possibly many were possessed of sufficient capital to assure eventual success if they had been trained in its judicious use, but, on the other hand, many were settled on inadequate blocks, or were engaged in the wrong type of cultivation, or both.

The bank management, through the direct liaison of bank inspectors, was faced with the responsibility of making farmers out of this mixed bag; and one can imagine the strain now put on the Narrogin inspector. There were not only more approved applicants for loans, but also an increase in the variety of loans requiring supervision. There was clearing to be inspected and measured, fence lines to be laid out, dam sites to be selected; advice on machinery repairs and the erection of required sheds and buildings.

The railway system continued to expand,[13] and in Mitchell's district, as in all others, the initial announcements or promises of construction precipitated a rush to select land, and subsequent inevitable requests for Agricultural Bank assistance to develop it.

His work and responsibility was further increased by his serving as local liaison officer for the Seed Wheat Board,[14] established by the Scaddan government in 1912 to supply farmers in the drought affected areas with seed wheat, fertilizers and fodder.

There were 1,600 applicants in the first year, 120,000 bushels being supplied, mostly through the state farms at Chapman and Narrogin. The balance of the seed wheat collected was of all breeds and kinds and doubtful quality, and was facetiously referred to as 'Scaddan's Glory'. Repayment was to be made over two years with security against the land, with a deed of assignment over the crop to ensure control over expenditure and equitable distribution of proceeds between creditors. The Board continued to function until the end of 1914, the Department of Agriculture acting as the administrative authority.

With the aid of the Seed Wheat Board and extension of credit facilities from the Agricultural Bank, the distress of 1911 was tided over, but only to be revived and accentuated in 1914 by the state's greatest single agricultural disaster till then: a drought of extreme severity over a very wide area. A couple of simple comparisons will serve to illustrate the harshness of the blow which then descended on the state's farming community. Of the 1,072,653 acres sown for wheat and hay in 1911, wheat accounted for 612,014 acres which yielded 4,258,904 bushels at 7.12 bushels to the acre. In 1914, 1,867,547 acres were sown for wheat and hay, but 1,367,012 acres stripped for wheat yielded only 2,624,190 bushels at a disastrous 1.9 bushels to the acre. What this meant to the inspectors on the spot in the devastated areas—among them Charles Mitchell in Narrogin—can only be guessed.

However, despite the derogatory remarks made about the seed wheat still being supplied, the 1915-16 season harvest was a normal one, and the odium of Scaddan's Glory was soon forgotten.

In September 1914 it became evident that a great number of farmers in the wheatbelt urgently needed assistance to enable them to remain on their farms and continue operations: storekeepers in all districts were unable to assist further. Anticipating legislative authority, the government (W. D. Johnston, M.L.A., Minister for Lands) decided to form the Farmers Assistance Board, consisting of W. Paterson, managing trustee of the Agricultural Bank, and G. L. Sutton, Agricultural Commissioner for the Wheatbelt, to provide deserving cases with stores. Assistance with stores was quickly followed by requests for seed wheat, fertilizer and fodder. Local chaff was provided by the Farmers Assistance Board, and other fodders such as lucerne, maize, bran and pollard were imported from the eastern states of Australia, from Java and the Argentine. Also, action was taken early in 1915 to enable distressed farmers to agist their horses in areas where feed was available.

Into this hotch-potch of political wrangle and economic tangle regarding the insolvent and the needy, the 'worthy' and the 'unworthy', was born the Industries Assistance Board Act, assented to in parliament on 2 March 1915. This act was born in a great deal of internal controversy as to its propriety, and it is doubtful if, in the history of the Western Australian parliament, any other act so exceeded the original intention of its creators.

Early in the year it was recognized that the administrative work of the Board was becoming exceedingly heavy, and on 24 February 1915 J. P. Camm, district surveyor of Northam, was appointed to assist the other members of the Board. The Board's work continued to increase and correspondence was becoming enormous, so on 14 July 1915 the Board was reconstructed as a separate departmental entity to give it more muscle and flexibility. C. G. Morris (Senior Assistant of the Under Secretary for Lands) was appointed chairman, with A. M. Oliphant and J. P. Camm in place of Paterson and Sutton, whose other commitments precluded their devoting the necessary time to Board business. Office accommodation was provided on the third floor of the A.M.P. Society's building in Perth and arrangements were made for the Board's accounts to be kept separately.

Designed as an expedient to meet the gravest set of circumstances the wheatbelt had ever encountered, the Industries Assistance Board Act was expected to solve the wheat industry's problems in one year and then tidy up the few loose ends.[15]

The new act was the antithesis of the Agricultural Bank Act, which lent long term for improvements against the security of the land. Overnight the government had moved into the normal preserve of the trading banks and merchants: the provision of seasonal requirements and day-to-day needs of the farmer against the security of crops yet to be grown. Without any reflection on the hard-working officers with their multifarious duties, the Board had no chance of success.

For the protection of settlers and to maintain the Board's security it was decided by the government to declare a moratorium for the assistance of settlers. A Royal Commission was therefore appointed under the Postponement of Debts Act by Executive Council on 9 November 1915, and on its advice a proclamation was issued on 10 November 1915, postponing debts then due or about to fall due: this relief was designed for settlers and other persons to whom aid had already been given—or might later have been given—under the Industries Assistance Act.

Operated entirely by Perth-based staff recruited largely from government departments, with little or no knowledge of farming conditions, the overcentralized system was unable, in its first year, to ensure that seed wheat, superphosphate and other supplies were available to all farmers in time for the 1915-16 sowing. The resultant chaos was one of the reasons behind the resignation of John Scaddan's Labor government on 27 July 1916.

Frank Wilson's National government took over the reins of office, and the Industries Assistance Board Act was renewed for a further year. To promote a closer and more personal interest on its part, the government amended the act (No. 3 of 1916-17), placing the administration under the Minister for Industries in lieu of the Colonial Treasurer. The position of Minister for Industries in this new government was held by James Mitchell (27 July 1916 to 28 June 1917). The apostle of optimism, he coined a timely epigram that confused his critics and heartened his supporters: 'There are no bad seasons, it's just that some are better than others.' What could you do with such a leader but follow?

Eventually the new government concluded that the Industries Assistance Board would continue. At the same time, the government realized that all aid to farmers from state funds, for whatever purpose, should be allotted from the same administrative source. Hitherto, for several years in the past, aid had been given in parallel: from the Bank on long term for permanent improvements and bill of sale for stock and plant, and from the Board for carry-on and day-to-day requirements under a blanket cover which took precedence over the registered securities. In August 1916 the Agricultural Bank and the Industries Assistance Board staff, while still accommodated in separate buildings, were brought under one administration—the Agricultural Bank.

W. Paterson, one of the original members of the Farmers Assistance Board, was appointed general manager; E. A. McLarty became assistant general manager, G. W. Simpson was appointed acting manager on 1 August 1916 and manager on 1 July 1917; C. Treadgold became secretary and chief correspondence clerk; and W. Beck was appointed accountant.

As an indication of the chaotic situation of the time—one in which nobody wanted to pay, intended to pay, or could pay—the Board took over from the Agricultural Department the Seed Wheat Ledger, which showed outstanding accounts totalling £40,000.

The Royal Commission on Agricultural Industries was appointed by the Governor, Major General Sir Harry Barron, on 13 September 1916.[16] There was no place for recriminations or 'witch hunting'. Mitchell, as Minister for Lands, had initiated the policy of wholesale selection of land by settlers of whom a high proportion were either moneyless or generally unsuitable, and this policy was primarily responsible for the position which faced Mitchell as Minister for Industries. He displayed rare perception and administrative ability. Decentralized control with sympathetic and knowledgeable staff was the answer, and who better than the Agricultural Bank field staff was qualified to undertake this liaison?

The Agricultural Bank administration decided to divide the wheatbelt into districts under district inspectors, and to institute a scheme of monthly visits to farms for on-the-spot payment in accordance with work carried out. Twenty-two additional inspectors were appointed to give effect to the scheme, and district offices were provided at Perth, Geraldton, Northam, Kununoppin, Kellerberrin, Beverley, Narrogin, Bruce Rock and Katanning. It was subsequently found advisable to divide the last into Katanning and Ongerup.

Charles Mitchell was one of the number appointed as district inspector, remaining stationed at Narrogin. Three inspectors were chosen by Head Office for his district—Lawrence Reid, G. R. A. Shepherd and Thomas Winslow Box. Mitchell was instructed to try to find a fourth one. The further histories of Reid and Shepherd in relation to the Bank have been lost and it is possible that, in due course, they reverted to employment within the Lands Department. Tom Box was appointed to Wagin in September 1916, and remained there till his death, in his sixty-third year, on 13 June 1943. Perhaps prompted by a premonition of things to come, Head Office provided Charles with an office—the rear room of the court house, with entrance from Fortune Street[17]—and on 1 March 1916 F. W. (Fred) Harrison, recruited from Perth, was appointed clerical assistant to the district inspector.

Later in the year, on 21 September, a district office of the Agricultural Bank was opened at Beverley,[18] with Frederick Henry Bostock as clerk-in-charge, co-ordinating the work of Inspectors H. S. Trotman,[19] stationed at Yealering, G. H. Birt[20] and F. E. Hitchens. These inspectors took over the northern portion of Mitchell's territory and the southern section of the area normally administered from Northam.

However, the establishment of these offices and increase in staff was not, by any means, the complete answer. The District Office was responsible for correspondence, records and interviews only. Ledger accounts were retained in Head Office, as was the prerogative of all decision making, to Mitchell a time-wasting and frustrating procedure; but, all in all, he was relieved of a good deal of hack work, and the line of communication between Perth and Narrogin was considerably improved.

Up to this time Mitchell, by reason of his 'short cuts', had hardly endeared himself to central administration. On receiving very lengthy, wordy letters from Head Office, he would study them carefully but would often merely write on the bottom: 'O.K.—C.D.M.' or 'I don't agree—C.D.M.', and return the annotated original. If further information was required, or later reference made to the original letter, Head Office was obliged to return the whole of the correspondence or—horror of horrors—the complete file, for Mitchell's information. The situation was relieved in short time by Fred Harrison's arrival in Narrogin. Fred was a touch typist with a classical education, and the perfectly phrased and beautifully typed letters he wrote on Mitchell's behalf soon became the envy of Mitchell's Head Office critics. In the ineffable fashion of administrators, Fred was snatched to higher places: considered too good for 'the bush', he was transferred to Perth in April 1917.

It was increasingly apparent that the Industries Assistance Board Act had come to stay—for some years at least. A majority of those farmers assisted had come to regard the provision of the act as 'a matter of right, not of privilege'.[21]

Official records of the Agricultural Bank relate:

> The 1918-19 season was a comparatively bad one, brightened somewhat by the adoption of better farming methods by the majority of clients. Despite the season a number of clearances were given, while in other cases settlers were able to improve their financial position, but the greater proportion showed little progress. Stoppage of assistance had to be resorted to in many cases where it was considered the settler had no prospect of redeeming his position—many of these settlers have remained on their holdings: they will be closely watched and every encouragement given to them where possible. It was decided by the previous Minister [the Honourable R. T. Robinson][22] that clearances from the Board were to be insisted upon as soon as the settler could pay his debts and handle sufficient to carry on. The Hon. Mr. Mitchell,[23] on the other hand, definitely instructed that clearances are only to be given where requested. It may seem strange but fully 50 per cent of the recent clearances under the late Minister's instructions had to be forced, as the settler desired to remain under the Board's supervision and control. Since the change of policy many requests have been received from settlers for reinstatement, but have been refused unless dictated by very special reasons.

The farmers' attitudes were certainly not the least of the pressures seeking the continuance of the Industries Assistance Board Act. The Board had, in fact, by then become a clearing house for other government departments, and all farmers' liabilities to the state—land tax, water rates, land rents, road board rates, and supplies from the State Implement Works—were now concentrated under its supervision. It was a perfect example of robbing Peter to pay Paul. One does not wonder that the influences pressing to continue the Board were many and varied.

An amount of £317,728 9s 7d was paid during 1917 on account of government departments, and during 1918, £93,800 was paid to government departments for land rents, bank interest and water rates. Owing to the poor season the Board could not see its way clear to make available the amount that had been provided for this purpose in the previous year, but each case was dealt with on its merits.

The Royal Commission on Agricultural Industries, among other directives, investigated and reported on the working of the Industries Assistance Board and remarked that the Board had undoubtedly saved the wheatbelt, but that in doing so, 'it had been guilty of every sin in the agricultural and administrative calendar'. The main conclusion of the Commission was that the Board had been unable to exhibit either economy or efficiency.

Neither were these harsh criticisms confined to the Board. The Agricultural Department was castigated for its failure to render assistance or guidance to inexperienced settlers. In the latter context perhaps one should note that with all the offices he had been called on to fill, the Agricultural Commissioner for the Wheatbelt, G. L. Sutton,[24] could hardly be expected to carry the whole industry on his shoulders; nor was it possible to obtain qualified staff to meet the unprecedented expansion. Neither could criticism be levelled wholly at the members of the Industries Assistance Board, which now comprised A. R. Richardson (chairman) and F. V. Cooke (part-time commissioner of the Agricultural Bank), and Robert Maitland Leake as the farmers' representative.[25] These men possessed long experience in agricultural conditions, but the concept of the act had envisaged neither projection beyond one year, nor the influences dedicated, as it were, to its continuance. The winding up would now take time.

The Royal Commission visited the Narrogin district early in 1917, but Mitchell did not give evidence before it. Why? is the intriguing and recurring question. Was he not asked, did he decline or did he neglect to volunteer?

Evidence was given by two of the inspectors under Mitchell's control, however: L. Reid[26] at Toolibin on 13 February 1917, and G. R. A. Shepherd[27] at Wickepin three days later. No record exists as to whether their evidence was volunteered or ordered, but one can conjecture, at least, the degree to which their depositions might have been formed by consultation with their immediate chief who, for any or all of the reasons suggested above, might not have felt it proper to attend the Commission's sitting and testify in person. It is an interesting observation that Inspectors G. H. Birt and T. W. Box, who were to continue in employment under Mitchell until either retirement or death claimed them and who came to be regarded as the hard core of his team, did not give evidence. Was it a case of not rushing in 'where angels fear to tread' or was there a sense of futility?

To Mitchell the ways of administration were distant and confusing; neither to his understanding nor liking. Here was a man with a high degree of natural intelligence, possibly the most experienced field officer in the employ of the Bank, a fearless fighter for the rights of others, certainly not inarticulate on behalf of others, yet apparently neglecting an opportunity to advance his own and the district's interests. This attitude, completely out of character, keeps repeating itself. Did loyalty or respect for his superiors outweigh his undoubted moral courage? Was it his place to criticize or suggest amendments to policies and procedures which his brother James had either initiated or opposed? Would his praise indicate he was a tool of his brother? Would his criticism be quoted politically? Did the discipline of his army service prompt his apparent action of performing duties without question?

THE APPOINTMENT OF A CLERICAL ASSISTANT to relieve Mitchell of the ever-increasing and detested paper work generated action which he had been contemplating for some time—the purchase of a Model T Ford car. With the ever-increasing demand for loans and inspections, and the continual travel this brought about, it had become increasingly apparent that the horse-and-buggy days were gone. Roads were being extended; time was now of the essence. Mitchell found that it was necessary also to work harder than ever. Not only did he have his own district, that area west of the Great Southern Railway line, to control as an inspector, but he also had to find time for supervision and training of both office and field staff.

Now more than ever did it devolve on him in all conscience to ensure that his staff had the application, knowledge and patience to sort out those who had an aptitude for farming and had some hope of eventual success. There must be no mistakes of judgement—people's futures were at stake. He was never at home. He knew farming success lay in a way of life, a slow painstaking result of individualism, plus muscle and a lifetime of

Caroline Mitchell

William Bedford Mitchell

Charles Mitchell, 1906

Ida Marian Mitchell, 1906

Agricultural Bank staff, 1912

Front row, l. to r., V. L. East, H. Paterson, W. Beck, C. A. Campbell, H. M. Dallas, H. M. Rose, G. A. Clark; *second row,* seated, W. Grogan, E. A. McLarty, A. R. Richardson, W. Paterson, F. V. Cooke, A. G. Hewby, G. M. Richardson; *third row,* Miss Durham, W. H. Cooke, A. C. Morrell, H. J. Robinson, J. B. Taylor, J. F. Cowcher, sweat. He knew that the application of these attributes, supported by reasonable financial help and technical advice, could make the farms, build the homesteads, the townships, the saleyards, the country pubs, their sports grounds and race courses. He knew their world would be hand-hewn, rough, but lasting; sustained with wives, families and neighbours.

Mitchell recognized also that the country towns must play a part in the development of community spirit because they were the meeting place of those engaged in farming. If a good spirit existed in a country town, it would do much to draw the farmers together and develop progressive movements. If, however, a country town was merely a place where things were bought and sold with a maximum of hard bargaining and a minimum of mutual friendship, it would be most unlikely that the result would be helpful in the advancement of either the town or the farming industry surrounding it.

There was no limit to the encouragement he gave to farmers—who in those days inclined to self-help and co-operation, as opposed to today's 'gimme' attitude—but it took a lot out of him. He wrote his mother in June 1917, from the Darkan district, that he was tired of his routine trips and of being away from home, and wished he could give it up. However, 'I suppose I will have to keep going a bit longer, nothing else better offering', he concluded.[28]

This was his lowest recorded ebb, but one can understand his becoming dissatisfied at times, and especially at this particular time. It was the middle of winter, it had rained continuously for three days, and it would be at least a week before he would get home. Then he had to write his own reports and deal with those of the staff. The strain on him was terrific; and there appeared to be neither end nor future.

His feeling of hopelessness was reflected in the trustees' Annual Report for 1917:

> A series of erratic seasons, absence of commercial credit, scarcity of labour and ravages of rabbits have combined to make the position of the farmer one of extreme difficulty. There is no overflow of population to create a demand for land and the Trustees can see little hope of improvement while War conditions continue.[29]

However, there was one bright spot. Having been given authority to recruit a field inspector at local level, if a suitable person could be found, Mitchell promptly gave the chance to his close friend, Harry Chipper, whose father had died some three years earlier leaving a farm hardly big enough to support Harry's young and growing family. Chipper had the qualifications required in a field inspector—a rural background, farming knowledge and a good education.

He often related a story indicating the degree of confidence Mitchell reposed in his staff, and his undeviating policy of allowing them to develop their own initiatives.

On 23 September 1916 he accompanied Mitchell in the Williams area to assist and gain experience. Ida was expecting another baby, and Mitchell had arranged for any news to be telephoned to the Williams Hotel. They enquired shortly after lunch: there was none, and it being a cold wet day, and the company and atmosphere being most congenial, they lingered. Then came the news: a boy—the first boy. Mitchell's joy knew no bounds; everyone had to share in his good fortune. Congratulations were exchanged and numerous toasts drunk. Time for departure had by now been long delayed, but when finally it came:

'Come on Chipper', said Mitchell, 'Time we left. You drive.'

'I've never driven in my life', protested Harry.

'There's nothing to it', replied Mitchell. 'Here's your chance to learn, I'll show you.'

Starting the Ford and pointing it in the right direction, he turned the wheel over to Harry and promptly went to sleep on his shoulder. Harry would swear he got to Narrogin with little trouble, but then couldn't wake the sleeper to show him how to stop it. He had to drive round and round till he ran out of petrol.

The son who was the primary cause of this unusual homecoming was subsequently christened 'Philip': 'lover of horses'. He was Ida's special gift to her beloved husband, a son and heir who, she prayed, would follow in Charles's footsteps in a love of which she herself had never been jealous—a love of horses.

WESTERN AUSTRALIA'S LONG, DRY, SUMMER MONTHS made a tinder box of the vast uncleared and grassed areas of the Narrogin district. Mitchell recognized early that community effort was the most efficient answer to the threat of fire to life and property. He not only preached and organized 'control', but rigorously attended every fire, and on such occasions his experience and his acceptance by the farming community made him a natural leader.

When Mitchell first came to Narrogin, means of communication were sparse, and reports of fires often came from observers who had identified the location as a matter of 'guesstimate'. Before the installation of the bell at the town fire brigade station, built in 1926, telephone advice was to the railway station, where there was always at least one engine working in the shunting yards. When news of a fire came in, the driver would sound three sharp blasts on its whistle, repeating them at minute intervals until it became obvious that the message had been received and action had been taken. George Grainger, one of the few old identities remaining, recalls that although he was among the group of employees who invariably attended fire calls on working days, he and Mitchell were regarded as also being available each weekend as neither of them regularly played any sport. One called for the other and away they went, first by sulky and later by car or truck. Their respective families recognized the 'must' of the attendance, and that their husbands and fathers would be home again when they arrived.

Fire-fighting equipment at first consisted either of a long-handled shovel or a bough torn from nearby scrub. There was no let-up once you arrived at the scene and there were never enough volunteers to relieve anyone needing a 'breather'. They worked in the heat and breathed smoke continuously, and this contributed most to exhaustion; and it was just such a set of circumstances which cost Mitchell the death of Harry Chipper, his close friend and trustworthy associate.

Mitchell responded to a fire call on the East Wickepin road in March 1919, and took Harry along with him. All day long they toiled without respite, and on the way home took a dip in a dam to refresh themselves. Harry collapsed in Mitchell's arms—the effort of fire-fighting, and perhaps the shock of the cold water, burst an already inflamed appendix. Having rushed him to the nearest hospital (Nurse Kelliher's 'Vailima' on the eastern side of the town), Mitchell had then to assist at the emergency operation conducted by Doctor Mackie and Nurse Kelliher. Harry rallied but septicaemia developed, necessitating a further operation two days later; he passed away on 11 March.

Mitchell never recovered from Harry Chipper's death; his family and closest acquaintances felt the mourning was life-long. Whether it was taking Chipper to the fire, brushing aside any suggestion of his feeling unwell; whether it was personal involvement at the hospital, or inability to accept fate, one can only surmise. As he sometimes remarked to his family: 'How could such a perfect specimen of manhood die like that?' Whatever the reason, Harry's widow Maud and her young family became Mitchell's self-appointed care, a responsibility which never ceased, although it diminished with the years. Mitchell asked Harry's brother-in-law, William John Ness, to take the vacant job. Ness agreed; he had been looking after three small family grazing properties, doing some horse dealing and cartage contracting. There was some criticism of the appointment, but Mitchell could have repeated, with equal sincerity, the expressions of Sir John Forrest, Premier in the state's first parliament, when accused of giving preferment to friends and relations in respect of appointments. 'Why shouldn't I? Whom do I know better, whom do I trust more?'

Ness, as far as is known, was the only inspector who did all his work on horseback, mostly in the Wickepin district. His favourite horse was called 'Inspector' and was apparently of unlimited capacity. After carrying Bill for six days a week he was borrowed during weekends by Lieutenant Stan Loftus who, among his many civic contributions, was Commanding Officer of the local troop of the 10th Light Horse.

On 20 March 1917 Marshall Fox was re-appointed divisional surveyor, Narrogin, with a composite office staff. He re-opened the Narrogin office in the clerk-of-courts building, and Mitchell was forced to find other accommodation.

Despite this unintentional unkindness, Fox was delighted to find Mitchell still in Narrogin. With his tremendous knowledge of the Narrogin district he had in the past been able to save Fox time-wasting classification reconnaissances before he surveyed new areas for settlement. Mitchell had never let him down, and now—with and through his field staff—could assist him even more.

C. W. (Charlie) Edwick was the next inspector appointed—to Kulin—following his discharge from the Army in early 1919. Edwick welcomed the appointment with Mitchell as his boss, having met him previously while farm labouring in 1913 at Ajana, when Mitchell was acting as official guide to a group of Agricultural Bank senior officials, comprising A. G. Hewby, W. Grogan, E. A. McLarty and W. Beck.[30] This group had scouted the country from Bruce Rock to Northampton, and were returning to Bruce Rock, paralleling their northern reconnaissance some 50 miles to the west. They travelled in one single-seater buggy with a long tray for stores and camping equipment, and two two-seater buggies, each drawn by two horses. Edwick recalls that he had been particularly attracted to Mitchell as a superior because of his wide knowledge of horses,[31] and his appreciation of farmers' efforts, which he was always prepared to acknowledge and reward.

Mitchell, following his usual custom, spent the first two weeks with the newly appointed inspector. One of the first farms they visited was occupied by an elderly married couple who, to give their adopted son an opportunity, had left a lucrative position in the commercial world of Perth some years before and selected a virgin property near Kulin. Although they had worked hard and intelligently, they were beset with bad luck, and had lost their entire team of horses from poisoned water.

'Find a new team for them, Edwick, out of those we're holding on abandoned farms, and keep the values to a nominal figure', Mitchell instructed.

'What about mortgage documents to cover the increased debt?' enquired Edwick.

'You just get the horses and advise my office of details', Mitchell said. 'It's Head Office's worry after that. These people have a real and urgent need.'

Ten horses were promptly selected and supplied at an average of £5 per head. Edwick had 'got the message'.

The following day Edwick and Mitchell went to observe a group of four blocks which had been taken up by Perth business men some seven years before, and on which Bank advances had been made for fencing and ringbarking. Care of the assets had not been maintained nor, in three cases, had the owners ever been resident, not even the minimum as required by the Land Act.

'Don't butt your head against the granite of this crowd's political influence', Mitchell advised. 'I've tried—there's nothing you can do. But don't let them treat you as putty.'

As they approached one of the farms in question, Mitchell gave another instance of his insight into the character of the people among whom his job lay.

'I'll bet the lazy so-and-so who's living here as their representative will have the gall to ask us in for a cup of tea, and leave his wife to turn the handle of the winnower on her own', he remarked, as they approached the homestead. And as Edwick recalls, such was the case.

Lewis Edmund Whitford was the next inspector to be appointed, on 17 November 1919, following his discharge from the Army. He was Mitchell's choice, having farmed in the Narrogin district prior to enlistment, and being one who needed minimum instruction. Whitford was stationed at Harrismith and remained there till his death in 1947.

The Bank, on quitting the clerk-of-courts building, leased a shop,[32] between H. Brown (butcher) and Illingworth (grocer) adjoining the court house on the western boundary. Harrison left in April 1917, and was replaced by A. B. Webb, who occupied the position of clerk-in-charge until his resignation in early 1921.[33]

The railways continued to expand,[34] with Mitchell's responsibilities increasing in like proportion. Edgar (Tinny) Everett was appointed clerical assistant on 8 April 1918, and became in addition Charles's close friend and confidant.[35]

S. R. (Stan) Loftus, a local boy, was employed as junior clerk in November 1919. Mitchell, with a natural eye for contours and farm lay-out, spurred on with an ebullience that would not be satisfied till his 'dream' of a Greater Sports Ground came true, had drawn many sketches with a stick on the ground, or on backs of envelopes, and had painted even more pictures in words. It was from these 'mud maps' that Loftus, living close to his boss, came to perceive the ideal which he transcribed into the plan which won the Australia-wide prize of £25 for designing the lay-out of the Narrogin Greater Sports Ground.

On the death of William Paterson in 1920, E. A. McLarty, his deputy, was appointed managing trustee, with W. Grogan moving into the position of deputy and A. G. Hewby to that of manager. Mitchell was offered the position of chief inspector, but as acceptance would have entailed state-wide travelling, with Perth as his headquarters, he declined the promotion. Few knew his reasons; the family doctor, D. W. H. Mackie, had advised Mitchell privately that in view of his wife's health following the birth of their second and youngest son Charles on 6 February 1919, he should never be more than a day's travel from Narrogin and within an hour's reach of a telephone.

E. G. Kelso was appointed acting chief inspector on 1 April 1920, and confirmed in the position twelve months later.

1. The predecessor of the present Liberal party. The term 'Liberal' was first applied to Sir Ross McLarty's government, which assumed office on 1 April 1947.
2. Known also as Eastern Goldfields Water Supply Scheme and Goldfields Water Supply Scheme. It was under the control of the Goldfields Water Supply Administration until 1912 when it came under the Public Works Department.
3. Records held by the Department of Agriculture.
4. From Bureau of Census and Statistics:

Year	Conditional purchase holdings & reserves	Area (million acres)
1908-09	4,527	1.662
1909-10	5,168	2.609
1910-11	5,696	3.120
1911-12	5,141	2.997
1912-13	2,569	1.376
1913-14	1,298	611
	24,399	12.375

5.

Year	Area under Wheat (acres)	Yield (bushels)	Average
1906	250,283	2,758,567	11.0
1908	258,011	2,460,823	8.6
1910	581,862	5,897,540	10.1
1912	786,183	9,125,245	11.6

Source: J. S. Battye, *The Cyclopaedia of Western Australia* (Adelaide, 1913), vol. 2, p. 140.

6. Bruce W. Leake, *Reminiscences* (Perth, 1961).
7. These were: G. M. Richardson, A. C. Morrell, H. J. Robinson, J. B. Taylor, J. E. Cowcher, G. A. Becker, J. Mills, A. G. White, H. M. Dallas, and W. G. Sinclair.
8. Née Eliza Agnes Farr. Her father, Edward Farr, was the first settler in the district. Their property was known as 'Wogolin Gate'.
9. Reserve No. 22247 Williams Location 9952 of 47 acres.
10. Except for two short intervals, Johnston served in the Legislative Assembly from 3 October 1911 to 3 October 1928. He was elected to the federal Senate as from 1 July 1929 and continued a senator until his death on 6 September 1942.
11. Assistance available from the Board administering the Industries Assistance Act, normally available only to those farming prior to 1917.
12. The eighth parliament, under the Labour Premier, the Honourable John Scaddan, with the Honourable T. H. Bath as Minister for Lands, was opened on 1 November 1911. It was the first stable Labour government in the state and carried out much of its platform policies in the next five years.

13. Extension to Dumbleyung-Kukerin were opened on 3 May 1912; to Wickepin-Corrigin on 8 June 1914; to Yilliminning-Kondinin on 15 March 1915.
14. The members of the Seed Wheat Board were: G. L. Sutton (Agricultural Commissioner for the Wheatbelt) as chairman, with William Paterson (managing trustee of the Agricultural Bank) and A. Berkley (Department of Agriculture and Industries).
15. The I.A.B. Act was amended annually up till 1953, to continue the act for a further twelve months, when it was re-enacted for five years, at the end of which period the time limit was deleted.
16. Members of the Commission: John Owen Giles (chairman), Frank Evans Venn, Henry Harold Paynter and Bernard Lionel Clarkson.
17. Narrogin Lots 39 and 40 were reserved for the purpose of an Agricultural Hall, Reserve No. 5740 being gazetted on 12 May 1898. Erection of the hall was made possible by a grant of £400 from the government and public subscription of £160. In March 1907 the government purchased the building for £1,000 for use as a Crown Lands office. The proceeds were utilized towards the building of the existing Town Hall and Municipal Offices in the following year. The Rural and Industries Bank occupied the premises in October 1945 and purchased the freehold in 1953. Ross East as clerk-of-courts and State Savings Bank agent, shared occupancy with Mitchell.
18. Closed on 24 February 1918.
19. The subject of Eleanor Smith's prize-winning biography, *The Beckoning West* (Perth, 1966).
20. George H. Birt was subsequently transferred to Narrogin district and stationed at Kukerin, where he remained until March 1923, being relieved by Inspector Lance Bird. After three years at Lake Grace he was transferred to Newdegate where he remained until his retirement in 1937 (aged sixty-seven years). Mitchell described Birt as the perfect example of a field officer.
21. The Royal Commission on Agricultural Industries in its *First Report* (22 November 1917), item 83, 'Credit for Agriculture', reported: 'Although settlers universally give the Agricultural Bank all praise for its assistance, there is an unmistakable tendency to regard further State advance not as a privilege but as a right and we fear that the best results will never be obtained while our settlers rest content with the idea that it is the State and not the individual who must be relied on to promote progress.'
22. Minister for Industries in the Lefroy ministry, 28 June 1917 to 17 April 1919.
23. Mitchell succeeded Robinson as Minister for Lands in the Colebatch ministry.
24. George Lowe Sutton, born 23 October 1872, was appointed Agricultural Commissioner for the Wheatbelt on 18 May 1911, which

position he occupied until his appointment as Director of Agriculture on 1 July 1921. Retired on 23 October 1937.

25. Appointments for all three dated from 22 August 1916. Alexander Robert Richardson, grazier and farmer, was elected to the Legislative Council for the North Districts in June 1887. He was elected to the first parliament, 1890, as the member for De Grey, and appointed Commissioner of Crown Lands in the Forrest government, 4 December 1894. He resigned office 13 March 1897.

    F. V. Cooke was one of the most successful farmers in the Northam district.

    Robert Maitland Leake, farmer, of Kellerberrin, later operated also in connection with the Soldiers' Settlement Board in the repatriation of World War I ex-servicemen.

    Richardson carried on as chairman until his resignation 30 September 1923; F. V. Cooke till his death 13 August 1931; and R. M. Leake resigned 30 April 1921.

26. As extracted from the minutes of evidence given: Laurence Reid, described as inspector, Industries Assistance Board and Agricultural Bank; length of service from September 1916.
27. George Robert Arthur Shepherd, described as inspector, Lands Department and Industries Assistance Board; been in the district seven years and employed by I.A.B. since September 1916.
28. Letter from Charles Mitchell to his mother, 23 June 1917, in the possession of Mrs Nora Rae.
29. The trustees' report amplified the conditions that had begun to prove too much for Mitchell: 'Properties reverted, following the harvest failure of 1914, totalled 876, plus a carry over of 34 from the previous year. 227 of these properties had been sold, leaving 683 on hand at 31st December 1917. 980 enlistments of Bank Clients had been noted. In a considerable number of these no satisfactory arrangements had been made for maintenance of improvements. To minimize the wastage on untenanted securities cropping leases had been arranged where possible; 132 leases were current at 31st December 1917.'

    No record is held as to what proportion Mitchell had of these responsibilities, but one can be sure he had his share.

30. Occupying the respective positions of chief inspector, manager, deputy managing trustee, bookkeeper and accountant.
31. Edwick's knowledge of horses, aided by four years with the 10th Light Horse, was equally extensive. He was recognized as the 'local vet' for the fourteen and a half years he was at Kulin.
32. Portion of Lot 1216, of 8 perches, occupied today by chemist Brian Twaddle.
33. Webb, born in 1887, was a man of wide experience and above average ability, and gratuitously assisted Mitchell in the formation of the Greater Sports Ground. A talented artist, he did a coloured

caricature of 'Ginger the Nob' which was publicly displayed until after the peace celebrations for World War I. He resigned to join the Perth Technical College staff as an assistant Art master.

34. Extensions to Kukerin-Lake Grace were opened on 25 November 1916; to Bokal-Wagin on 20 November 1917; to Bowelling-Bokel on 10 December 1918.

35. Everett, born 1884, was appointed to the permanent staff on 1 October 1925, and died on 21 August 1932 from a gun-shot wound, self inflicted. Mitchell felt his loss very keenly.

9

Narrogin: Private Life 1908-1921

On arrival in Narrogin the Mitchells lived for a short time in an old house in Fairway Street facing the Great Southern Flour Mill. The exact location is unknown. From there they moved to Charlie May's house, 8 Falcon Street—always referred to by the family as 'Morack'—where Charles fitted up an office on the eastern end of the front verandah. The vacant block adjoining the office contained horse yards and stables and a shelter for the house cow.

With a practical eye to the future, he travelled to Perth in 1909 to enquire of Head Office if it was safe for him to purchase a home in Narrogin, and was advised most strongly against it—his future location being even then under review. In two subsequent approaches he received the same advice. Twenty-five years later he was still in Narrogin and then, and not until then, did he acquire his own home—only to be transferred in the following year to Northam.

The remainder of the family were born in Morack: Nancy Caroline ('Caroline' after the paternal grandmother) on 14 August 1909; Gertrude Kathleen ('Gertrude' after mother's sister and 'Kathleen' a close friend), never referred to by any name other than 'Peggy', on 10 June 1912; Philip Rossiter on 23 September 1916; and Charles Rossiter on 10 February 1919.

During these early years, their social activities were meagre—there was little time and less spare cash. Mitchell was usually away from home for three weeks at a time, and all Ida would know for certain was that he was 'out east'—there were few post offices or telephones to make contact. After a spell at home for a week or ten days he would be off again.

The one recreation Mitchell could permit himself was to observe and enjoy horses—in action and otherwise. He loved them, understood them; they shared a love equal in its own way to that he bestowed on his family. One knew by his demeanour how much a new horse intrigued him—hands thrust into pockets, pipe in mouth, he moved slowly around the animal, studying both good and bad points. The deeper the hands thrust, the more interested he had become. This was one of the few times the pipe was forgotten and went unlit.

In the days when the horse represented both transport and traction, to be recognized as 'a good judge of horseflesh' gave Mitchell immediate entry into any society. He was an authority and accepted as such, his opinion sought and valued.

There is only one known incident of his ever having mistreated or overdriven a horse. One evening when he was camped alongside the road beyond Yealering, a farmer passed him. He had been on the move for some time, with no communication between him and Narrogin, and when the farmer commiserated with him about 'the accident' he was shocked and apprehensive.

'What accident?' he demanded.

'Your daughter's', the farmer replied, with equal surprise. 'They took her to hospital.'

Mitchell was 50 miles from home, but his magnificent team deposited him at the kitchen door by midnight. Ida was as surprised to see him as he had been to hear about the accident. Nancy had cut a finger off and was now sleeping peacefully and in no danger whatever: the Mitchell family was news over a large stretch of the Great Southern and Eastern wheatbelt, and news travels with remarkable swiftness in country districts. Mitchell decided to stay at home since he was there, and while his phlegmatic wife put on the kettle for a cup of tea, he went out to look to the horses which had served him well. It had been a false alarm, but their great hearts would have served him equally well in a real emergency.

At the time of Mitchell's arrival in Narrogin the Race Club, stationed between the Yilliminning Road and the railway line,[1] had run into financial difficulties, and was virtually in recess. Interest in racing was being maintained and strengthened through the Narrogin Picnic Race Club,[2] which held its meetings in a paddock which belonged to 'Micky' Brown.

Micky, born Michael Brown in Tullarone County, Kilkenny, Ireland, in 1850, came to Western Australia in 1857, when the family migrated. Following the dissolution of a farming partnership with a brother at Wandering while still in his teens, he purchased his father's interest in the grazing run at Noombling.

Travelling widely while searching for sandalwood, he perceived that Narrogin appeared the logical linking point for the two ends of the Great Southern Railway then being built simultaneously from Albany and Beverley. He backed his judgement with the purchase of 200 acres in freehold, and in 1888 built an inn (the first licensed premises on the Great Southern) near the site of the present Municipal Sale Yards. He married Elizabeth White for love, not, as was sometimes hinted, for the practical opportunity of then employing her father, Joseph, to tend the several acres of grape vines planted adjacent to the inn, and to act as winemaker—a form of industry which proved a valuable adjunct to his trade.

On the W.A. Land Company's later surveying a townsite adjoining the railway station, he purchased a block, built the first brick structure of any kind along the line and transferred his licence to the Hordern Hotel—named after the promoter of the Great Southern Railway. He became Narrogin's largest land holder, and pioneered the Jitarning district. He had a family of twelve children, was a patron of racing, and was invariably the owner of an outstanding stallion.

Despite the disparity in ages, Mitchell and Micky became close friends. The Brown home, 'Carabin', a large stone building on the corner of Federal Street and Clayton Road, was one of the few Mitchell visited frequently. Horses, farms, and the future of Narrogin were the principal topics of conversation.

In the paddock, Williams Location 1136, on the eastern side of Fitt Road (now Narrakine Road), adjoining the present golf links to the south, there were outstanding clumps of luxuriant red gums offering perfect shade for picnic parties. The paddock formed a natural amphitheatre, with the course in the centre. The riders crossed the creek in the southern section over a plank bridge covered with sand, but in the north galloped through the creek bed.

In his eighty-fourth year Frank Fitt recalled quite vividly the 'race paddock' where he first rode at the age of thirteen, which places its existence before the turn of the century. Maintaining his weight under seven stone with ease, Frank rode regularly and widely at country meetings, until the age of twenty-six, when his mother forbade him to ride again because he had accepted a ride on Micky Brown's bay stallion, Duke of York—a savage brute at the best of times.

Two years later Frank married Elizabeth Margaret, daughter of Harry Finnegan, first curator of the Narrogin Sports Ground, who had come from Kalgoorlie to work for Tom Hardie.

James Fitt, Frank's father, who owned farming land adjoining the north-west corner of the 'course', had purchased an aged, sway-backed, long-necked, well-bred chestnut gelding named 'Strathmore' from Tommy Boyle of Brookton ('the biggest, ugliest horse you ever did see', is Frank's description) and also leased two horses—History and Gregory—from Tom Hardie, which he raced with only moderate success. Gregory was eventually sold to Mitchell as a lead horse with Nemo, and was re-christened Nominus.

Nominus must have retained quite a turn of speed. Mitchell used to boast that when coming on a newly settled couple at Kulin, who had not tasted fresh meat in months, he promptly saddled Nominus and ran down a kangaroo for them on the nearby plain.

Walter Fitt, a younger brother by two years, also still alive and active, was of sturdier build and rode only occasionally, but recalls winning a Cuballing Cup on Strathmore, beating Frank who finished second—Walter still has a suspicion that Frank 'pulled' his mount to let the 'old man's' horse win. In 1917 Walter married May White, a niece of Micky Brown's wife.

There was a grave amongst the grove of red gums, unmarked except for the six by three mound. Frank remembers his father's saying, 'it was some teamster'. The grave was treated with due reverence—the children might run over it but the adults never used it for a picnic seat. There is no trace of it today—wind erosion and cultivation have completely effaced the mound.

Meetings were held at Micky Brown's at every opportunity, the races interspersed with picnic and gymkhana events. The crowd would begin to arrive by 10 a.m., and never left till dark—fun and games were the order of the day. Emmet Brown, seventh and youngest son of Micky Brown, recalls his first memory of Mitchell at a race meeting in 1912. Mitchell had entered in a melon race: contestants had to mount and ride to a certain point, dismount, and grasp a piemelon under each arm. They then had to remount as well as possible, and, guiding the horse with knees or teeth, gallop back to the starting point. Mitchell, in his exuberance, mounted so fast that he went straight over the horse, and knocked himself out. There he lay, unconscious, eyes shut but still grasping his melons. Emmet, wondering whether he was dead, was immensely relieved when he opened his eyes and enquired, 'Where's the whisky?' From then on he was Emmet's hero. The hero image was further enhanced when in 1915 Emmet had the opportunity of accompanying Mitchell on a trip to the Kondinin district, where Micky Brown's father had a pastoral lease of some hundred thousand acres which provided excellent winter grazing for sheep. A railway linking Narembeen to Merredin was projected,[3] and the government was gradually resuming all the pastoral leases for agricultural occupation. Mitchell was to report on compensation claims for improvements effected.

They camped near Karlgarin Hill. Dingoes, numerous and to all intents hungry, surrounded the camp fire. Emmet swears they formed a complete circle; all he could see was their eyes gleaming in the dark. Mitchell made no comment, and appeared quite unconcerned. Emmet never closed his eyes, but Mitchell slept soundly, although he woke occasionally to throw wood on the fire. As the fire blazed, so the circle of eyes retreated; as it died down they came closer. 'In the morning he made absolutely no mention of the dingoes', Emmet recalls. 'He gave me the feeling that he believed I had slept all night, just as unconcernedly as he had, although he must have known I was scared stiff. Of course, I never let on, either!'

Mitchell teased his friends unmercifully, but never would he knowingly hurt them.

In later years a young clerical officer obtained Mitchell's permission to accompany him on a trip to the bush during his annual leave. 'All you need bring is your blankets', Mitchell advised him. On the first evening, closely observing the boss's procedure, the youngster made a hollow for his hips and with his boots for a pillow, settled down on the other side of the fire. Sleep was impossible. To ease the cramp and discomfort he arose, at least half hourly, to replenish the fire or gather wood. On the next day Mitchell made a point of driving through some teatree scrub, and casually mentioned that many bushmen made a 'mattress' from the leafy branches. By coincidence they camped that night alongside a teatree thicket and for an hour the young officer picked and piled branches till he had a pile 4 feet high—then slept the clock round. Not by as much as a twitch of an eyebrow did Mitchell indicate that there was anything incongruous with either night's procedure.

PRIOR TO WORLD WAR I, MITCHELL DECLINED the appointment of Justice of the Peace for the Williams-Narrogin district. Conscious of the honour and its attendant responsibilities, he yet would not be classed with some contemporary recipients whose appointments savoured highly of political favours. Charles was his own man, no-one could own him; he owed allegiance only to his Maker, his Sovereign and the Bank.

What his politics were no-one ever knew, or whether he voted for a party or the man. The only remarks his closest friends can recall as to his political views referred only to his brother, Sir James. 'I don't care for what Jim is doing', or 'Jim is barking up the wrong tree', or 'Jim seems to have the right idea this time'. He had adopted, naturally, the Public Service tradition of silence under public criticism, which is the service's necessary expression of loyalty to its political leadership. It was not only in the fields of politics that he had convictions and unwavering principles. Born and raised in the faith of the Church of England, he rarely attended church services after leaving school, possibly only to be married; yet he was neither atheist, agnostic nor irreligious. Ida had his full support in the attendance and education of their children in their Church. He neither discussed nor criticized organized religions: he judged each man's beliefs on their manifestations in his conduct.

There were some who have described him as a humanist—a most abstract term, and one he would never have applied to himself. While he loved humanity, he reserved to himself the non-acceptance or dislike of the individual. There were others close to him who described him as a free-thinker. This was equally inappropriate. While he appeared to give to everyone, including his family, the right to choose for themselves as to religious observance, he allowed himself no latitude. He lived by the 'golden rule'—at all times a Christian gentleman. He believed implicitly that men were born equal and that within themselves they possessed the capacity for success or failure, in this life, and with reward or justice either in this or the next world. Adhering to this principle, he eschewed all groups or societies which gave members special privileges or influence. He considered it was as immoral to use such benefits to one's own advantage as it was to deny use of them to the less favoured.

Ambitious to achieve eventual independence through farming, rather than merely to supplement his income, Mitchell sponsored the purchase of Williams Location 8698, adjoining Traysurin

The Wattles
A recent photograph showing remains of Ida and Charles Mitchell's first home

Carnival for Belgian Relief Fund, 1915

Peace celebrations in Mackie Park, 1919

railway siding, in the name of Ida Marian Mitchell. The deal was made with William Cunningham Ramsay (whose chief land holdings were near Bunbury) as from 1 September 1916 for the consideration of £100 cash, and the taking over of an Agricultural Bank mortgage of £174. The property was then leased to and worked by the Mullan family. During the term of the lease further loans totalling £450 for clearing and fencing were arranged, the work being carried out by contractors. Mitchell supervised, and found additional costs from his own pocket. Local gossip has it that, with the aid of a bottle of whisky and a particularly tasty 'camp oven stew', he convinced the railway engineer that the proper place for a siding was adjacent to Location 8698.

On the declaration of World War I, Mitchell was among the first to volunteer; but was just as promptly manpowered by the Bank. He applied to join the 25th Light Horse, the successor of his old regiment, the Western Australian Mounted Infantry,[4] portion of 'B' squadron being attached to Narrogin.

Baulked in this wish, he would have liked very much to maintain an active association with the Citizen Force unit instead; but family obligations and his work left no spare time for the weekend parades and periodical camps.

His initial disappointment was lessened with the announcement that no A.I.F. mounted unit was required from Western Australia, but he was outraged when, later, following strong and persistent representations, 200 volunteers from the 25th were recruited as 'C' squadron of the New South Wales 7th Light Horse Regiment. By October 1914 volunteers and enquiries were of such number the 10th Light Horse Regiment was approved and Mitchell's friends flocked to join it, providing their own mounts.

As a palliative, Mitchell threw himself into every patriotic effort. At first his contribution was limited to his spare time, and there was little of this until after the opening of the Narrogin district office in 1916 when, through the appointment of additional field staff and the assistance of clerical staff, he had more time at home. During the early years of the war we read of him only as the starter for the Narrogin Picnic Race Club on 26 October 1914, in aid of the patriotic fund, and again at the Easter Picnic Race meeting in the following year for the Belgian Relief Fund.

Later he became a stalwart of the Ugly Men's Voluntary Workers of Western Australia, an organization which raised charity funds during and after the war. A friend, one with either a sense of humour or colour harmony, presented him with a green felt hat. When he first put it on, he said: 'I look quite a nob!'—and so he became 'Ginger the Nob'. Wearing the green hat came to signify that he was fund raising, with the catch cry *'A Bob for his Nob'*. His appearance and enthusiasm caught the public's imagination, and A. B. Webb, his clerical assistant, produced a coloured drawing, emphasizing the mop of red hair, which was placed in Mick Anderson's (barber and newsagent) window, and made the shop the rallying point of his friends. Prominently displayed, stuck with stamp paper to the inside of the window, was a set of verses, the authorship of which Mitchell credited to the clerk-of-courts, Ross East:

GINGER'S DAY

Every morning Redhead Ginger,
Ugly man of Narrogin,
Puts the hard word on the milkman
And requests him to put in
Just a Bob for my old Nob.

Local tradesman—even 'Peelers',
After breakfast are his game,
All part up a 'Deener'—smiling,
So that Ginge of evil fame
Scores a few more for his Nob.

Waylays Merchants, Bankers, Sawbones,
Sports and Parsons one and all,
Ginger slugs them for their silver,
They can't bluff him with a stall,
For begob its for his Nob.

Round the pubs he gaily prances,
Acids all the bums to vote,
Makes a hole in their finances,
Stops a crack but scores a note,
Twenty bob for my old Nob ! ! !

Puts the glad-eye on the flappers,
Even if his Missus' looking,
Blarney, bluff, kidstakes and googoos,
Sets him with his pencil Booking
Easy money for his Nob.

Tie and collar long discarded,
Hat lost somewhere in the mud,
As day closes Happy Ginger
Goes home with a collar stud
And a thousand for his Nob.

An extract from the *Great Southern Leader* (10 May 1918) shows him in action:

> The Narrogin Branch of the Red Cross ran an Ugly Man competition early in 1918 to raise funds. The outstanding feature of the effort was undoubtedly the energy infused into the Ugly Man competition, and in this connection the bulk of the credit is due to Mr. C. Mitchell, who entered as 'Ginger the Nob'. Mr. Mitchell made the pace a cracker from the start, putting some 'real ginger' into the fight, and was responsible for more than half the surprising total of £415 14s 5d realised by this means. The runner-up (Mr. Albert Kitchen) has to thank an energetic committee of young people for bringing him so close to carrying off this most coveted honour.

The letter Charles wrote, during the month following, to his mother, one of the two people to whom he could bare his soul, shows rather more of his character and thoughts:

> Ida wrote and told you all about the Ugly Man competition I won fairly easily. The people in Narrogin and districts were exceptionally nice and I got money from all parts to put on 'Ginger the Nob'. Some people spent in small amounts (the ones that are hard pushed) which I did appreciate, knowing it was for the cause. At the same time, I knew they wanted me to come out on top. I heard recently that Amy was a great worker for patriotic purposes; someone from Bunbury told me.
>
> I think we will have to do a lot more than what has been, for the men are badly needed, and I would like to go and do my bit. I am fit and well and when I see these boys going I feel down in the dumps for days afterwards. If I had sufficient money to provide for Ida and the children I would have to go. I do wish we had conscription, then a man would know where he was. I can't see how Ida could manage so have to give up the idea of thinking about it.[5]

The central executive of the Red Cross, in recognition of his personal efforts and example, voted to make Mitchell a suitable presentation. He objected to the proposition. Under no condition would he accept anything; he had only 'done his bit'. A gold wristlet watch was then presented to his wife, inscribed 'To the esteemed wife of "Ginger the Nob" 1914-18'.

The first welcome home to the district's ex-servicemen, in February 1919, took the form of a municipal concert, and one of the choice items was an original song dedicated to Mitchell and his confreres who, in their fund-raising efforts, could not and would not take 'No' for an answer:

On a recent Saturday I had really lost my way,
As I tried to find the entrance to the Club,
For the way was very long and the sun was very strong,
When I got thrown out of Bushy's pub.[6]
But, Oh, I never knew that Lavater[7] was in view,
Till he came and asked what trouble ailed me so,
As I dropped a silent tear, he said, 'What you want is beer',
And I couldn't, could I, very well say No.

When our Aussie boys in France, stopped the 'orrid Huns advance,
By their gallantry at Villers Bretonneux,
Fritzie got it in the neck, and his plans went all to wreck,
As he staggered under each successive blow.
He was belted every stunt, all along the shot-torn front,
Till he found that out of France he'd have to go,
Urged by bayonet, bomb and gun, his retreat became a run,
And he couldn't, could he, very well say No.

When Nick Anderson[8] is sent to his future punishment,
He'll meet Wharton,[9] East,[10] and Chipper[11] down below,
Hardie Bros[12] will be there, so will Charlie Mitchell's hair,
Making all the dismal place with fire aglow,
Jimmie Mann[13] and Butler[14] too, Jonah,[15] Baird[16] and quite a few,
Other chaps, as private lives as you all know,
Made the pace in this life, hot, in the next they'll go to pot,
But they couldn't, could they, very well say No.

Mitchell was Master of Ceremonies and paraded in his green hat, displaying also the largest red tie ever seen in the town.

With the background of a job well done, he was chosen logically as one of the organizers of Narrogin's peace celebrations, and became a member of the finance and general sports committee. The celebrations took place on Saturday 19 April 1919, in the form of a procession, football match, carnival and fancy dress competition. At 11 a.m. the procession commenced the day which continued uninterrupted except for the lunch break and five minutes for silence at noon, when the Last Post was sounded in honour of those who had paid the supreme sacrifice. These celebrations made history. Mitchell received the highest recommendation for his work and, to fill in his day, he was with Police Sergeant Martin Barry (later inspector), joint organizer of motor control.

Not to be outdone in the peace celebrations, the Farmers and Settlers Association held a sports carnival on 17 March, and in the evening engaged two halls, one for dancing and the other for pictures and musical entertainment; of course Mitchell was in the thick of it. His efforts were not all show; during these years he was a committee member of the Narrogin branch of the Australian Natives Association which, incidentally, conducted a most successful sports meeting on Monday 27 January 1919, raising well over £30 for patriotic purposes; and he was also a foundation member of the executive of the Narrogin branch of the Soldiers Repatriation Committee.

His willingness to assist anyone in trouble was equalled only by his impetuosity in rendering that help. Returning one day from Corrigin to Narrogin by car, he came across another car with its engine on fire and the owner trying vainly to put it out with handfuls of sand. Mitchell yelled: 'You want more sand than that. Leave it to me.' Grabbing an empty petrol tin, he cut the top out with an axe, managing at the same time to cut his left arm, with one of the tin's jagged edges, from wrist to elbow. This was not the moment to bother about such trifles, and by the time he had scraped up enough sand with his opened petrol tin to smother the fire, blood had saturated his suit. With sand adhering, as well as the grease he'd managed to rub against, the suit was ruined. Because of his help however, only minor damage was sustained by the car, and the owner, in his gratitude, offered to make good the damaged suit.

'Not on your life', said Mitchell, 'the insurance company will meet it. Look at what we saved them.'

The insurance company not only turned the claim down, but wrote that their business was insuring cars, not paying anyone to act as fireman. Mitchell hit the roof. 'I'm eleven quid out of pocket', he roared. 'Next time I see a car on fire, I'll ask which company it's insured with, and if it's the —— company, I'll throw petrol over it.'

The long-widowed Caroline Mitchell died at 'The Willows' on 9 August 1918, from double pneumonia. A matriarch, loved and respected by all, she well deserved the compliment, remembered with humility and pride by Nora: 'A wonderful family, they never did a mean thing.'[17]

The *Bunbury Herald* stated in its obituary notice:

> Yesterday afternoon quite a link between the present and the past was severed when Mrs. Caroline Mitchell, of 'The Willows', passed away in her seventy-first year, a victim of the dread scourge of pneumonia. The deceased lady had a sweet and gracious disposition and her undoubted charm was felt by all who came into contact with her not only of late years, but in the days gone by, for she may almost be spoken of as one of the pioneers of this district and played her part

> in the days when the foundation was laid of the position which we are in today.
>
> For years Mrs. Mitchell has led a somewhat retired existence, but she previously more than earned her gentle dalliance upon the borders of life in the evening of her days.[18]

A fundamental forthrightness, neither vicious nor malicious, was the birthright of the entire Mitchell family. Forthright remarks would bubble out at unpredictable and even inopportune times. This trait existed even in the family circle, but here, once it was 'off the chest', it was completely forgotten. Caroline's forthrightness, lightened with a generosity of spirit and sense of humour, had in no small measure contributed to the inheritance of these attributes in her children.

One day Amy, in great excitement, drew her attention to some sailors who were leaning over their side fence stealing apples, advising mother 'to send for the police'. 'Send for the police, indeed', replied Caroline. 'The poor men may have been at sea a long time and starved for fresh fruit!'—and then proceeded to invite the sailors to fill their pockets.

Her chicken broth was famous and freely available to patients at the nearby hospital, whether they were known to her or not. Dr I. J. Flynn, Commonwealth Medical Officer, and one of the town's two private medical practitioners, spoke of her as his 'best unpaid nurse', and had no hesitation in frequently calling on her dietary assistance.

True to her promise, 'Kelly's' farm was bequeathed to her sons, Harry and Charles, in equal shares. Charles was not concerned: he had relinquished his interest in it in 1908, and Harry would send him a cheque when he got around to it, or could afford it. Harry, generous of mind and heart, took no action; his younger brother would contact him in due course and let him know what value he put on his share.

Following the birth of baby Charles Rossiter in 1919, Ida's health deteriorated. Doctor's bills piled up and it became apparent that the prospect of going farming as a future family venture was more remote than ever. His wife must come first so, on February 1920, he sold the Traysurin property to Frank Herbert Walker of Dudinin for £1,000, including the debt owing to the Agricultural Bank.

Although he had pressing need for money, Mitchell would not appeal to Harry for his share of Kelly's. He would have considered his own need personal and private and, in addition, any such request might financially embarrass his brother. However, although the two beneficiaries could see no reason to concern themselves, this attitude did not suit Jim, Harry's eldest son. On finishing school in 1920, he continually pressed his father to purchase Uncle Charles's share and ensure the family ownership of the property which had now been named 'Uda Bremer'—'Uda' being the native name given to Harry during his surveying days, meaning 'big head', and 'Bremer' the romantically based family name. Harry and Jim travelled to Narrogin to see Charles—and Jim confessed to a salutary lesson in 'gentlemen's agreements' while there.

'Uncle Charles couldn't have been more fair', he said. 'In less than twenty minutes he had calculated his interest at £1,200 [at 1908 values], and five minutes later had agreed to contribute £500 of his bequest towards an annuity for spinster sisters Kate and Amy.'

Trouble started when the Taxation Department refused to accept adjudication of Charles's half interest at £1,200, contending legally that stamp duty should be assessed on the value of the half share as at date of execution (1920). Mitchell set off for Perth to see the deputy commissioner himself: if it was good enough for Mitchell to accept a moral issue then it was good enough for the department to recognize it also. The argument was long and acrimonious but the transfer was accepted at 1908 values.

1. Dumberring A. A. Lot 164 was gazetted Reserve No. 7417 on 7 December 1900, and leased to the Race Club. Dumberring A. A. Lot 112 was added to the Reserve on 17 October 1901, as creeks in the northern portion of Lot 164 did not permit full development of the race course.
2. Formed in 1909, with Dr J. B. Lewis as president and R. (Ted) Edwards, secretary and handicapper.
3. The Kondinin-Narembeen extension was opened on 22 December 1917, and the Narembeen-Merredin on 16 March 1925.
4. The Western Australian Mounted Infantry formed prior to federation continued under the existing title as a volunteer regiment until 1903 when it was reformed as a militia regiment, the 18th Australian Light Horse Regiment (W.A.M.I.). In 1912, following the introduction of compulsory part-time training, the regiment was redesignated 25th Light Horse (W.A.M.I.).
5. Letter from Charles Mitchell to his mother, 23 June 1917, in the possession of Mrs Nora Rae.

6. N. Bushalla's Duke of York Hotel.
7. George Lavater, secretary, Narrogin Road Board.
8. Barber and newsagent.
9. H. Wharton, local carrier, father of Dick (later manager of Elder Smith & Co., Narrogin).
10. Ross East—clerk-of-courts.
11. Harry Chipper, Mitchell's special pal.
12. Tom and Ted Hardie.
13. Manager of Millar's Timber and Trading Coy.
14. Ernie Butler, jeweller and watchmaker.
15. T. J. Jones, manager, National Bank.
16. H. C. Baird, manager, Bank of Australasia.
17. Attributed to the Misses Alderson (deceased); overheard and repeated by Mrs R. J. Anketell.
18. *Bunbury Herald* 10 August 1918.

10

The Administrator 1922-1929

On the opening of the Narrogin branch of the Agricultural Bank on 13 March 1922, Charles Mitchell, by virtue of his position as district inspector, assumed administrative control. His status as such would have been equivalent to that of country branch manager of any trading bank. Tradition and Public Service restriction governed the terminology.

I could not consider other than that Mitchell, in his own humble way, felt that in this appointment he had accomplished both a personal ambition and something in the family tradition. Elder brother James, now Premier, had been a successful bank manager. Both brothers Will and Tom had joined the Western Australian Bank, although Will resigned from managership of the Dongara branch after his marriage to Fanny Priscilla Pearse on 12 May 1898, to go farming. Nora always considered him the brightest and most high-spirited of all the boys, and one who would have gone far in the banking profession. An erstwhile Western Australian Bank officer, long since retired, who knew the younger members of the Mitchell family very well, was stationed at Dongara in the mid 1920s. He recalls his amusement, reading the record of plaintive appeals to Perth Office, to learn that 'Mr. W. B. Mitchell will neither call at this office as requested, nor answer correspondence. What shall we do?' Will must have had a poor opinion of the generation of bankers which succeeded him.

Tom was manager of the Marble Bar branch at the outbreak of World War I, and resigned in December 1914 to become one of the reinforcements to the 10th Light Horse regiment. He stayed on in England after the Armistice, attached to Army administration, till the end of 1919. He rejoined the Bank in January 1920, and resigned in August 1933, also to go farming.

The first branch established under the decentralization programme, so urgently sponsored by Sir James as Premier, was opened at Northam (on 8 January 1922) with Alban Clarence Morrell as district inspector and Alexander Henderson as clerk-in-charge. The experiment was not the success envisaged; frictions occurred, partly through lack of initiative and understanding at branch level, and partly through the reluctance of a section of Head Office administration to co-operate and assist.

The Premier fumed; decentralization had to work, and quickly. In his planning he had the active support of A. M. T. (Tracy) Gordon, chief audit inspector, and A. Berkeley, state Treasury inspector. They moved together and Narrogin was selected as the next branch, with temporary offices in the Town Hall Chambers.

The successful functioning of the Narrogin branch was a must. The initial staff gave the appearance of special selection. W. H. (Wally) Cover, with two months practical experience as paying officer at Northam branch, was appointed clerk-in-charge.[1]

The branch carried on in the two front rooms of the Council Chambers, severely cramped for space, until late in 1923 when it was transferred to the cottage in Egerton Street which had been vacated by Mitchell and family on their moving to 'Culbina', which had been made available to them through the good offices of Ted Hardie.

Mitchell's cup of happiness brimmed over. He had a position of responsibility, coupled with authority—more vaunted than real—but this had still to come home to him. He had a six-roomed house, four of stone and two of weatherboard, on a farm within close reach of town. Ida and the five children loved the open-air life—four were of school age, the baby, 'Bricky', being only four years old.

The branch remained in the stone cottage, still cramped for space, till the end of 1924, when it moved into the old court house building[2] at the corner of Egerton and Earl Streets (opposite the fire station)—this remaining as headquarters until taken over by the Rural and Industries Bank in 1945.

Surviving members of the original staff have vivid memories of their tribulations following the opening. Isolated and devoid of specific instructions as to procedures and branch administration, consolidation and the techniques evolved were mostly 'by guess and by God'. Guessing rightly more often than not, and aided by the Providence who helps those who help themselves, the branch was an immediate and continuing success. The first auditors' report—obviously a success—was presented in the presence of the Premier, Sir James Mitchell, and was followed by the opening of six new branches—Katanning, Kellerberrin, Geraldton, Bruce Rock, Kununoppin and Bunbury.

The early 1920s witnessed a rising plane of prosperity; loan moneys were freely available and concentrated on the spread of primary industry. Wheat farming began to turn from a way of life to one of investment: 'a quick profit and get out'.

This period saw also the trend towards mechanization, with a proliferation of both tractors and trucks. Roads were bad, however, and as maintenance service was generally very distant and not yet skilled, neither was as yet an economic proposition to the average farmer.

Into this maelstrom of change, demand, high pressure salesmanship and extended credit, Mitchell was pitched head first. Never one to swim with the stream, nor pass the buck to Head Office, he made his own decisions, wrote his own policy. He had the respect and confidence of the managing trustee and his co-trustees, but he saw too little of them. He did not have their ears as did Head Office's second echelon. Several of these he considered trivial, pinpricking 'silvertails' who exercised too much authority without judgement, and told them so to their faces in no uncertain terms. The consequent rift was wide, and filled with constant friction which was never wholly healed. Undoubtedly Mitchell did himself harm, but he could not compromise on terms which to him would have seemed deceitful. He could be blamed for his uncompromising standards, which might be construed as idiosyncracies, but as often as he, like all men, stood at moral crossroads, not one of his detractors ever made the accusation that he took the wrong turning for the sake of profit or position. His selflessness was not the posturing of religious scruples nor the publicity-seeking of politics.

As the district inspector he was neither deliberately pig-headed nor obtuse; but he was, in his own opinion, being continually overlooked and neglected. He wanted to do more, he earnestly wanted authority to see his points of view, but how to accomplish these things escaped him. On the one hand there were the distant proponents of the traditional laws, in whose view the laws of administrative experience, which they alone interpreted, provided unquestioned answers. On the other hand there was Mitchell, the man-on-the-spot, possibly with a God-given insight, subscribing to those laws of reason which adapt themselves to meet the times, to provide norms and guidelines of human action, and which influence men who follow them to act reasonably and morally.

Members of Mitchell's staff still pay him credit. He did not presume on his position, and he did not interfere. He possessed the ability to convey confidence, to delegate authority without undue committal of responsibility. He and he alone was always responsible for decision. Members of his staff were given every encouragement, and while he was kindness itself in rectifying and explaining their mistakes, he was their fiery champion should Head Office or client or official ever speak disparagingly of any of them. He neither ranted nor explained at undue length; if he ever erred, it was in brevity. It was taken for granted that the person addressed was his equal in intelligence: all one needed to do was listen and remember, and even if the moment of truth was delayed, it always came in time.

To Mitchell, paltriness was unforgiveable; like public bad manners, it could claim no excuse. Woe betide his staff member who displayed this weakness. He was tolerant of most things, except dishonesty and snobbery, yet at the same time he was chary of undue familiarity. Despite his innate friendliness, he never addressed any of his staff by christian name. 'Mister' or 'Miss' were part of his address, and anyone summoned by surname only knew he was accepted.

The direct approach was part of Mitchell's nature, part of his way of life; his brusqueness, a pose. Kindliness was inherent and this virtue was the key to his tact and courtesy. Devoid of guile, he did not look for it—and certainly detested it—in others. His staff came to refer to him as 'C.D.M.', and over twenty years gave him their complete loyalty and effort. Few managers enjoyed the love and respect which seemed to be C.D.M.'s natural due, but relations never descended to the modern level of first-name, back-slapping, pub-mateyness. The 'Christmas party' is an example of the simple and forthright way in which he operated. Each Christmas Eve at 4.30 p.m. all adult male members of the staff were invited to the Duke of York Hotel for a drink. C.D.M. led the conversation, bought three rounds and then excused himself.

One of Mitchell's precepts in customer relationship was often: *If you dislike someone, stay away from him. Otherwise you'll find yourself upset, then getting wild and spoiling your judgement.*

That instruction to his staff, more often by example than precept, is exemplified by the memory of one of them that one afternoon there floated out of his office, in stentorian tones: 'Of course you're not getting a fair go—what do you expect coming to see me? You know what I think of you and I've repeatedly told you to stay out of my sight. Write me a letter, then you might have a chance!'

One of the few criticisms levelled against Mitchell, invariably told jocularly but just as invariably implying deprecation, was that he never passed up an excuse or opportunity to quit the office, to view a horse or consider somebody's problem, even if it were not necessarily related to bank business, so long as it was out of town. While the criticism is valid enough, nobody who ever saw Mitchell in action as he dealt with the multitudinous long- and short-term problems which crowded onto his desk, would deny him the relief which he gained from his brief forays away from them. I know: I worked with him for years, and there were times when he would be frustrated and baffled to the point of incoherency. These were the times when he turned, instinctively, to the open spaces in which he had been reared and in which his very character had been formed—in which he still found release from the tensions of the desk-man he had, of necessity, become.

'I'm going to see so-and-so', he would announce, or 'I'm going to look at a horse!'—and go he would, and he would invariably come back refreshed and ready to cope. The hours he spent away from his desk were often far more valuable than those he spent at it—and anyone who knew that, knew him.

Few of Mitchell's field staff are available for interview, but J. S. (Stuart) Garland, who was appointed a field inspector in 1928, recorded his impressions of Mitchell for me. Stuart received his initial training under Mitchell in the classification of blocks, made available for selection by the Lands Department, north and north-east of Lake Grace. 'A man with a rough exterior, but with a very kind and understanding nature', he wrote.

> The bulk of his vast district was then under the Industries Assistance Board and settlers were for the most part, poor and struggling. They were dependent for everything on the Board. Among the settlers there was, of course, the usual quota of drones and schemers. It was not always easy to sort this type out from the genuine battlers, and it was in this capacity that Charlie excelled. He helped in every way with the latter, but he was down on the former, and being the possessor of the real Australian vocabulary, he was never at a loss to make his point clear. They were very difficult times for both the settlers and the Agricultural Bank and its field officers. Charlie had the task of training his field staff, by no means any easy task. He was also a staunch backstop and supporter of his own staff, a very essential characteristic in those times, when practically every settler had a

grouch of some kind, and would write direct to District Office or to his parliamentary representative. But in no case would they be dealt with until Charlie had conferred with the field officer concerned. He was a very fair man with a great knowledge of his work, and it was a pleasure to have him as a superior officer and help in those most difficult times.[3]

In writing of those days one cannot escape the necessity of commenting on one of the most salient features of the situation—the general attitude of farmers between 1912 and 1929 towards meeting their obligations, and particularly their obligations to the Agricultural Bank; however, the approach should be on psychological rather than on moral grounds.

With the 1912 amendment of the Agricultural Bank Act, it was no longer necessary for the farmer to provide an equity, advances of 100 per cent for work done being permitted.

Possibly the majority of settlers were genuine in their desire to be farmers, but almost invariably they had little personal capital and less practical experience, particularly in management. The problems of the Agricultural Bank were not only added to with each new client but, in the light of experience, they were multiplied. Credit funds, when they became available to the farmer from sale of various farm products, were lodged with banking institutions other than the mortgagee bank; the machinery and superphosphate firms gave practically unlimited credit, the associated banks made unsecured advances against 'proceeds in sight'. Under such a set-up, the creditors mentioned above invariably secured priority in the distribution of seasonal income—the Agricultural Bank was fortunate if it collected annual interest charges. This procedure eventually became standard practice, and it was advice openly and freely given to Agricultural Bank clients—generally by opposition bankers who controlled the credit account and wanted to keep it in credit and merchants who wanted payment in preference to mortgage obligations—'only pay your interest, they'll never kick you off—the government will never permit it'.

The faults were not all on the farmer's side. Farms generally had not been well designed for the type of agriculture in which they were engaged, and with no limit on advances, many properties eventually became overcapitalized. However, these two aspects were made manifest in retrospect rather than recognized at the time of which we write. It is unfortunate that dishonesty had become a matter of necessity, but Charles Mitchell, of all people, understood the position. As a matter of course, or of conscience, he assumed responsibility to ensure to the best of his ability that, in his district, no farmer was exploited; and in return, that no farmer should exploit the Agricultural Bank or any of its various agencies.

Mitchell understood that the husband and father, the natural provider for his family, must be able to fulfil this function to complete his role and maintain his dignity. He believed that any attack on a man's pride raised a barrier that made communication or agreement all the more difficult, if not impossible. When there was not enough money to meet all his obligations, his family had to come first. Many and varied were the subterfuges employed to divert to family requirements money which legally should have been applied to payment of bills or interest, the only difference between the fundamentally honest and the dishonest man being the degree to which the diversion and employment of such funds took place.

A lady called at Head Office one day, identified herself as wife of one of the Bank's clients, and asked for an advance of £10 for urgent medical attention.

The young clerk, observing the lady to be visibly pregnant, was reluctant to question her further, and passed her on to the accountant. He also showed diffidence and so she was passed on to the assistant general manager, W. (Bill) Grogan.

Grogan listened to her plea, and also to her explanation as to why it had not been possible for her to contact either their inspector or District Office on her trip to Perth. Accepting her excuse, he presented her with a cash order (the equivalent of a bank cheque), which the lady placed in her bag. She then told Grogan that he wasn't 'such a bad old so-and-so after all' and that she would like to do something for him in return. Modestly turning her back she extracted a pillow from the upper part of her dress and gravely handed it to Grogan.

Grogan knew when, and when not, to press his authority. He did not seek return of the cash order, and moreover it is interesting to note that the incident did not detract from his deep humanity, as the following story indicates.

On a subsequent occasion, his table became the final resting place of a file which had already been the rounds of the lower echelons without evoking a decision.

The case concerned a bank inspector who had refused to pay a client more than the single men's rate of sustenance, although the farmer possessed a de facto wife and three children. The complaints of the farmer and his de facto wife had been loud and many; something had to be done, but who to take the responsibility? Grogan's decision, in his own handwriting, read:

'Pay at full rates. We are administering the Agricultural Bank Act, not the Ten Commandments.'

In recounting these times, distinction must be made between the 'ordinary' Agricultural Bank client to whom the above remarks apply, and the soldier settler, to 5,213 of whom approximately £6,750,000 was advanced between 1918 and 1940.

The Discharged Soldiers Settlement Act, No. 9 of 1919, and its subsequent amendments, provided for settlement on the land of discharged soldiers, or their immediate dependents in the case of the serviceman's death, as also of munition workers who were employed on active service outside the Australian Commonwealth.

The act was introduced by the Lefroy government in 1918, and assented to on 3 January 1919, only a short time before James Mitchell became Minister for Lands, in the Colebatch government on 17 April 1919. As Mitchell continued in office in the short-lived Colebatch government, and subsequently in his own as Premier, he thus controlled the act—as Premier and Minister for Lands—for the next five years.

It was administered by a board of four members, which was called the Discharged Soldiers Land Settlement Board. The Board's duties, apart from adjudicating upon the purchase of existing leases, and the setting aside of certain Crown lands for soldier settlement, lay in three principal provisions. It was to investigate the qualifications of each soldier applicant as to farming ability, and if satisfied in this respect, to issue a certificate of qualification. It was to allot land open to soldier selection, and to adjudicate between applicants when there was more than one application for any specific piece of land. It was also to recommend to the Agricultural Bank what assistance should be given and what advances should be made.

It will thus be seen that, once again, under these provisions, in the state's first undertaking to rehabilitate ex-servicemen in a farming scheme, one group of officials (the Board) chose the successful applicants, and another (the Bank) was presented with the responsibility of making their selection work. When, one is constrained to wonder, would officialdom recognize the basic wisdom of permitting the Bank to find its own clientele, make its own judgements and rise or fall by them?

The overall plan intended that each ex-serviceman would receive a fully developed farm, earning an immediate income sufficient to cover personal needs, to recoup the normal expenses attached to cropping and to provide a margin sufficient to meet interest and repayments due to both the Agricultural Bank and the Lands Department.

Conceived by officialdom with the best of intentions, to avoid the crippling weakness of the Industries Assistance Board in being forced to find continuing credit funds, it yet failed through lack of first-hand appreciation of the individual farm's earning power. Possibly the errors were not entirely those of the state. The Commonwealth, never renowned for its generosity and inherently suspicious of the approach of the various states for funds, could have assured an economic level of production by making available sufficient money. As it was, the state planners were forced, through lack of finance, to 'cut their coat to fit the cloth'.

In return for extending 'full opportunity' to the ex-serviceman, the state protected itself through the far-reaching and all-embracing Section 7 of the amended Act, No. 45 of 1919:

> Moneys advanced by the Board or the Agricultural Bank, under the provision of this Act, with interest thereon, shall be secured by a mortgage upon the land acquired by the discharged soldier under the provisions of this Act, or upon any other land of the discharged soldier. All moneys so advanced with interest thereon shall also, until repaid, by force of this Act be and remain a charge on the crops and chattels of the discharged soldier to whom such advances are made, and which may be upon any land so mortgaged to the Board or the Agricultural Bank as aforesaid, and on all other livestock (including the progeny thereof) and chattels of such discharged soldier acquired under the Act.

This iniquitous section was described as giving the Bank ownership of everything the settler owned, including 'his wife's wedding ring'.

To hasten the rehabilitation process, the Discharged Soldiers Settlement Board purchased, developed, or part developed, individual properties or large estates, which were then subdivided. After allotment, the Agricultural Bank moved in to provide any additional improvements as may have been prescribed, plus finance for stock and plant. On the other hand it unfairly stigmatized all farmers, solvent or insolvent (although, one feels, this was not its primary purpose), by publishing in a red, hard-covered booklet, which was revised annually, the names of all S.S.S. (Soldiers Settlement Scheme) clients, listed alphabetically, with home address and branch at which the account was domiciled. This 'dossier' was distributed to all merchants and firms dealing in the sale of farm produce. This was done in order that they would be fully aware of whom among their customers was or was not subject to the Bank's statutory lien on their income.

It is interesting to conjecture that the Discharged Soldiers Settlement Act was not only a refinement of Sir John Forrest's Agricultural Lands Purchase Act of 1896, but a re-enactment of a process dating into the second century B.C. when various Roman Emperors sought the break-up of the Latifundia (broad farms) —the extensive estates acquired by Imperial aristocracy. Nero and Vespasian settled a considerable number of repatriated soldiers and marine troops in the declining cities of southern Italy. These measures proved fruitless, and from causes and pressures not inconsistent with the failure of Western Australia's first large-scale attempt at long-term closer settlement.[4] The identity of the World War I resettlement was lost long before similar and more successful settlement was undertaken subsequent to World War II.

No land settlement scheme has functioned successfully where credit moneys have been made available from more than one secured source. The main irritation to the scheme under discussion stemmed from the impractical manner in which repayment terms had been drawn. On the one side the Agricultural Bank required payment of interest and instalments on their loans; on the other, the Lands Department pressed for its dues—repayment of principal and interest, of the purchase price of the land and improvements existing thereon—when settlers' cash resources had been reduced to or beyond vanishing point, sometimes even before the property had been brought to its full earning capacity, and sometimes when the land could not produce sufficient to meet even the charges.

St Matthew relates in his Gospels that 'No man can serve two masters: for . . . he will hold to the one, and despise the other . . .'[5] Neither was this a case of 'render therefore unto Caesar the things which are Caesar's',[6] for Caesar in this case took multiple forms.

There was little or no liaison between the Lands Department and the Agricultural Bank, each being mindful of its own preserves. The act was watertight and inflexible; first call on income from farm produce was reserved to the mortgagees.

At local level Charles Mitchell's problem was to keep the soldier settler contented, to foster his progress and to protect his rights while pointing out to authority that the majority of the properties were overcapitalized and repayment requirement out of all proportion to the net earning capacity.

The overall effect on the morale of the soldier settler was disastrous. His other difficulties alone were acute enough, and the only feasible way of successfully out-manoeuvring existing policy was to acquire land—in his wife's name, for preference—and thereon to run a flock of sheep, however small in number, to supply the excuse of a separate income—whether or not part of it was derived from Scheme security.

Local stock firm managers recognized the subterfuge, but took no undue advantage of it. Mitchell's word was law. He never let the firms down and they stood by him and his settlers—supplying the farmers reasonable needs, and literally serving two masters. The executives of both the stock firms and the Scheme had a good idea of what was happening, but took no steps to alter it—the man-on-the-spot had to find the answer and take whatever kicks came along. These managers were, like Mitchell, big men in hard times. They swore by him; his index memory, his sense of proportion, his protection of the 'trier'.

A young officer who was at the time attached to one of Narrogin's stock firms (and who was later to become Mitchell's son-in-law) had a problem regarding several of his firm's clients. Who better to refer to than Mr Mitchell, one of the patrons of his football club, a man of recognized wisdom and experience? The opportunity occurred soon enough, during a chance meeting on a country road near Wickepin.

'Mr Mitchell, I have been looking after three of our accounts east of here, and over the last three years their flock numbers have shown no increase', the young man explained. 'There doesn't appear to be any reason for this, and I just can't get any sense out of the farmers themselves.'

'Don't let it concern you unduly, young fellow', Mitchell replied, 'so long as your firm is holding its own. If you want to find out where the sheep are going, there's little you can do about it except camp on the place—out in that district even the wheat jumps the fence!'

The nature of the problem and the situation is crystallized in another story told in Narrogin even today.

The Agricultural Bank having entered into possession of a farm, its owner came into Narrogin office to plead with Mitchell for one more try.

'You've got no chance', Mitchell said. 'I won't have a bar of you. You can't be trusted—in fact you're a damned thief.'

The ex-settler continued to plead, and when he played his final card, 'for his family's sake', Mitchell relented.

'Alright', he said, 'you can have another go—but I'm going to watch you like a hawk. Put one foot wrong and you're out for ever.'

Some two years later the man returned.

'Mr Mitchell, you needn't worry about me any longer', he said. 'I've had two good seasons, and can now afford to be honest.'

The moral of this story would apply to the greater percentage of farmers at that time.

RAILHEADS CONTINUED TO EXTEND INTO MITCHELL'S DISTRICT—Lake Grace-Newdegate and Dwarda-Narrogin were both opened in 1926. Each new opening either presaged or followed expanded settlement, and with expanded settlement came expanded demand for Agricultural Bank Loans. Although the district inspectors were required in a dozen places at once, they had still to recruit and train their field staff—and Mitchell at Narrogin was no exception.

By 1928 he had a field staff of nine. To train and supervise them, and to inculcate in them his required standards and interpretation of policy to fit local conditions took every minute he could spare, and in this respect he found invaluable the time he would spend with them on visits around the farms. Although his visits were eagerly awaited by the settlers—there was always a fractious horse which only he could handle, colts to be gelded, a job of shoeing or a prescription for any of the wide variety of complaints common to work horses—he would almost never stay overnight in their homes; his parties always carried camping equipment and invariably camped out while on the job—and it was on such nights, under the stars, in the open air, that he played teacher. It was around the camp fire of an evening that he really excelled—relaxing and expanding. Supper finished, he would stoke his pipe to capacity, and with hat askew and pulled well down, would walk backwards and forwards, stepping over the fire, talking and reminiscing, invariably managing to kick the billy over during his perambulations. This was his method of communicating, by stories, mostly against himself, the essence of which portrayed his philosophies: how to sift the no-hoper from the trier, how to differ between the schemer and the farmer who cheated from necessity; how best to advise; and, above all, how best to face up to the requirements of courage and confidence. This absorbed, he believed, progress must then follow.

To Mitchell, the triers in his district were human beings and were entitled to be treated as such. They were to be protected from bureaucracy, and sometimes from themselves, and under no circumstances was any of them to be forced into a position in which he would lose his dignity or self-respect.

In his camp-fire talks to his field officers he relied neither on rhetoric nor oratory to get his message across, but his language was colourful and expressive. Perhaps in some way he resembled the One who first taught in parables: his stories too were simple, and were spoken by a simple person. He might stretch the long bow or dress his stories up, but he never did it to his own advantage. The colouring was the gift of the natural raconteur—to make the story interesting, to emphasize a point or sometimes to chiack his friends. Those he liked most he ribbed most, and always to their faces: those he disliked, he neither spoke about nor frequented their company.

The tyro and the uninitiated might not have readily grasped the relationship between the fire-side performances and the seemingly casual conversations during journeys by car: nevertheless, they were all part of the whole. Mitchell, in his humility, could not dogmatize; he knew too well from experience, bitter at times, that the variations of approach to the solution of each farmer's problems were as numerous as the settlers themselves. How hard he tried to express his thoughts without posing as an oracle can only be guessed.

All one was called on to do was to listen, and those privileged to listen unconsciously absorbed the philosophies, remembered the experiences, and subsequently, from the recesses of memory, drew on them. Mitchell would never have been aware of the extent to which he would govern thinking in the field he knew best, and loved.

With the average trip extending from two to three weeks it was not possible to carry either sufficient bread or meat. Mitchell could make an excellent damper in his camp oven, and taught his staff to live off the land, to enjoy the flesh of rabbit, parrot, emu and kangaroo. With the addition of a knob of bacon, an onion and salt to any of these ingredients, he would set his camp oven on the coals in the morning and reveal in the evening an epicurean delight that many a hungry field officer can remember to this day. Always there was the side dish of potatoes baked in the ashes.

The secret of enjoying emu and kangaroo lay in the mental strength required to hang, for at least five days, the emu leg, hacked off with an axe, or the haunch of kangaroo, more neatly butchered, but with feathers or fur still attached. The afficionado, moreover, recognized the difference in quality between the bush-fed emu and the bird which had enjoyed at least some time on the farmers' grain paddocks.

I could never masticate or stomach a kangaroo steak done à la Mitchell, on the coals, with the ash knocked off by hitting it against a convenient stone or mallee root or even against the sole of his boot.

The same procedure was observed even with the beef steaks, which were supplied usually on the first day of the trip. Even then one's gorge rose with the first mouthful—to this day I believe Mitchell grilled steaks this way to test the tyro's reaction rather than add to the flavour.

However, when married and stationed at Newdegate some years later, I found this experience invaluable. There was not much beef to be had at that time in what was essentially a sheep-raising district, and nobody knows just how steak-hungry one can become on a perpetual diet of mutton. My wife and I achieved quite a reputation as hosts for our frequent weekend suppers of steak. Week-old emu steak, cut across the bone and slid off the leg in circular pieces and fried and served with gravy, passed adequately for 'rump steak', particularly so when served by lamplight. The dark colour was no longer perceptible. The source of supply was a matter of envy, not enquiry. Served with fried eggs and washed down with bottled beer, it was a Newdegate gourmet's delight! The eggs and the beer were always supplied by the guests as their contribution and at their insistence, at 7d per dozen and 1s 9d a bottle respectively.

Distaste for emu, kangaroo and rabbit was not the province of only the few and the fastidious. It extended far and wide over the entire wheatbelt where settlers, still not at the stage of development where they might kill from their own flocks, had been brought close to the dietetic bone. The districts of Warralakin and Warrachuppin, settled in the late 1920s, were no exception. For these two communities, drinking water was drawn from Geelakin Well, a water reserve, No. 1378, situated some miles to the north. An added attraction to the meat-hungry water carriers was the fact that the soak was frequented also by 'scrubber' cattle—beasts which were unbranded and so unidentifiable and fair game in anybody's language. They came, originally, from either or both of two properties: from Mt Jackson Station, many miles to the north, or from the lease of George Maddock, an early settler who had built a mud-bat home at Wilgoyne Hill, with a unique roof thatched with reeds from the nearby soak.

The Maddock lease extended over an area east of the Rabbit Proof Fence and south of the Mt Jackson boundaries. With the institution of closer settlement, Maddock's lease was cancelled and he was required to remove his stock: naturally there were some that could not be found and many were not worth finding, and it was from among these and their progeny that visitors to the Well were wont to kill their meat. The practice was to stop the truck about a quarter of a mile from the soak and reconnoitre on foot to spy any beast lingering in its precincts. The hunters always came prepared to do the whole job on the spot. After being shot, the animal would be dressed from a hand-made gambrel, and block and tackle, suspended to a nearby tree, the rope being tied to the back of the truck to raise the beast from the ground. Back at Warralakin and Warrachuppin, the carcass—dressed with varying degrees of skill—would be apportioned and distributed among the neighbours, to be eaten at the first opportunity; there was no refrigeration apart from the homely Coolgardie Cooler.

The story is still told of the occasion when Ted Dawson, local farmer and hard case, won praise throughout the community for what was credited with being an unusually well-grown and well-conditioned beast—as opposed to the old and stringy cows which were the normal wont. It was many months before the local farmers became aware of what had happened. Having made the long trip to the well, and found it destitute of cattle, Ted had determined not to return empty-handed, and so shot and butchered, and distributed among his friends, a lone donkey he had discovered temptingly loitering within rifle shot.

THE FIELD PARTIES NEEDED GOOD SUSTAINING FOOD, for the day's work consisted of viewing, at the closest possible quarters, the land and its problems—and as far as Mitchell was concerned, the only efficient way to do it was on foot. Accuracy was insisted on: guesswork was a cardinal sin and short cuts taboo. There were few who could outpace him or outlast him, and perhaps, even, it might have been undiplomatic to try. This was the part of his job he loved and did best. He was always reluctant to return to the desk, although he was fortunate that his office ran, as nearly as such a condition was possible under the circumstances, on oiled wheels.

Rules were made to be broken, even Mitchell's own rules; but by himself. He had a very soft spot for the somewhat rotund Inspector Lew Whitford, stationed at Harrismith. If, when he stayed with Whitford, there might be a back-log of measuring to do, on either fence lines or clearing, he would borrow a spring cart or a sulky. Then, carefully measuring the circumference of the wheel, he would tie a rag round the rim, and while he drove and took the compass bearings, Lew kept record and also counted the revolutions of the wheel. These were later translated into yards and chains.

This labour-saving device was not practicable in the case of small areas or unduly angular pieces of ground. Here they walked; Mitchell again read the compass and counted his paces. Whitford held the field book, scaled ruler and protractor, and plotted as he walked.

Absorbed one day in this joint exercise, Mitchell trod on the tail of a brown snake which, to show its indignation, struck upwards at the nearest thing—and knocked the field book from Whitford's hand. The two of them took off in a flash and as Mitchell related later, by the time they had gone 200 yards, 'that fat old Whitford was 10 yards in front!'

'Have you been bitten?' Mitchell asked.

'I don't think so', replied Whitford.

'Then, what in the hell did we run for?'

Back they went to pick up the book, and there were the indentations of the snake's fangs in the cover—it had struck between Whitford's outspread fingers. The book was preserved for many years as evidence of the 'snake yarn'.

WHEN JESSIE STEWART, THE SENIOR TYPIST, retired in 1928, no-one thought she could be adequately replaced. Her replacement, Thelma Lucy Flute,[7] was not only an accomplished stenographer, but she possessed also an unlimited maternal instinct. It wasn't long before she had organized Mitchell's office chores, and secretly he revelled in her control. There was enough instinct left over for the rest of the office; Thelma gathered them all under her wing, whether they liked it or not. She knew exactly how an office should function, and quite gently but firmly impressed her requirements on any 'new boy' who swam into her ken. I recall my first brush with her, after I had arrived in Narrogin and had been installed as correspondence and records clerk, at the fine and experienced age of twenty-four. I had dictated some letters to her—Shirley East, the junior typist was on leave—and on receiving back from her the meticulously typed-up copies, I noticed that some changes had been made.

'Thelma', I observed, 'this isn't what I said!'

'No', she replied, quite unruffled. 'But it's what you *should* have said!'

Mitchell hated dictating letters or writing lengthy decisions, possibly from a lack of fluency. When he returned from a country trip there would be an enormous heap of files requiring his personal attention, at the sight of which he would shy like an unbroken colt and literally gallop out of the room. Any excuse was good enough to postpone this particular chore, and I witnessed the little play of his entrapment so often that it survives in my mind rather in the form of a drama.

Usually by mid-afternoon of the third day after his return, Thelma would get him cornered.

'Come on, Mr Mitchell, we could get these files finished before five o'clock', she would say. With all lines of escape cut off and excuses exhausted, he would give in. 'Alright Miss Flute, bring your book in.'

SCENE I *Charles Douglas Mitchell dictating.*

(Mitchell, one side of the table, pipe going full blast; the other side Miss Flute, notebook open, pencil poised.)

FIRST FILE *He reads it carefully and then says*: 'Not on your life!' *Miss Flute makes note, takes file.*

SECOND FILE *A letter from Head Office*: 'What in the hell's the matter with them—do they think we've nothing better to do than answer damn fool questions for every so-and-so who has the authority to write us? We've already given them this information. Tell them to use their brains if they've got any.' *Miss Flute makes note, takes file.*

THIRD FILE 'Too ambitious, beyond his capacity. Tell him to break it down.'

Miss Flute makes note, takes file.

FOURTH FILE *A short study and then*: 'He must be bloody-well mad, or thinks we are. Tell him so, Miss Flute, and make it pretty strong. It's the only language he understands!' *and so on until the heap is exhausted and so is Mitchell!*

'Miss Flute, it's nearly five! I'm knocking off! I need a drink!'

SCENE II *Next day.*

All the files are back on Mitchell's table for his signature. He reads each carefully before signing. The results are exactly what he intended to convey:

> We regret to advise that we are unable to approve of your request of the —— inst.
>
> With reference to your letter of the ——, may we refer you to our report of the ——, which apparently has not yet been brought to your notice. It is felt that the information contained in the report referred to may serve your requirements; but should there be any further points on which you may require our advice, do not hesitate to write us.
>
> We desire to commend you for the enterprising programme set out in your recent communication. However, it is felt that it would be wiser for the present to restrict your cropping programme only to those areas which are vermin-proofed. According to our records these would approximate three hundred acres for wheat and perhaps the recropping of one hundred for oats. This letter will serve as authority for you to proceed on this programme, with finance provided for super, bags and cartage.
>
> The proposition as set out by you has been examined; but unfortunately from the information held at this office it would appear to be, at this juncture, neither practicable nor economic. We have passed a copy of your letter to our inspector . . . with the request that he call on you at the first available opportunity for further discussions and investigation.

Mitchell may have lacked fluency, but his judgement was seldom at fault.

While his work took him into many unmapped areas, there is no record of his ever having been lost. He was a splendid bushman, with the unerring instinct of a homing pigeon; but the incident of his once having gone missing for three days is well worth recording.

He was on his way out to Holt Rock in April 1933 to relieve Inspector Frank Collett, who was to be operated on for appendicitis. He travelled via Lake Grace and Lake Biddy, in answer to a plea from Harry Ward, farming south of Lake Carmody, 'to have a look at the illness of a valuable mare'. There was no chance of saving the mare, however. It was shot, and Mitchell helped to dig the hole and bury it.

A few miles beyond Ward's he met a government dogger (dingo trapper) named Herbert. Herbert's conveyance was a Harley Davidson motor-bike with a side-car box attachment. The outfit was notorious. While it was always possible to hear Herbert coming at least a mile off (his exhaust pipe had fallen off), one could, if the wind was favourable, detect the odour peculiar to his outfit—an effluvium which had been aptly described as a cross between an open septic tank and a slaughter house—several minutes before one heard his machine. Scalps and pelts in various stages of curing, plus baits and lures carried to distract the dingo's attention, all contributed to this unique oderiferous atmosphere.

Mitchell knew Herbert not only as an experienced dogger and a great cook, but also as the possessor of an amazing fund of stories. He was the type of companion who appealed to Mitchell and it being close to nightfall, they made a joint camp, Herbert keeping to leeward. He explained to Mitchell that he had found evidence that a number of dogs were using the extensive mallee and scrub plain on which they were camped as their breeding ground and hide-out for forays on nearby farms, but with neither defined pads nor dog-proof fencing to direct their movements, he was having little success in luring them to the traps.

Mitchell knew the answer, and volunteered to assist. At daybreak he returned to Harry Ward's, where he uncovered the mare's grave. He severed a hind quarter with an axe and, tying it to his rear bumper, commenced a series of wide sweeps which all centred on the site already selected by Herbert to lay his trap. He then watched Herbert lure the male dingos from the 'horse-leg' trail to the traps, using the only scent (obtained from the female dingo) which would overcome the dog's inherent sense of caution.

Enthused by trapping one dog from the first day's effort, they repeated the exercise for two more days until the whole of the flesh was frayed from the leg. When eventually he arrived at Holt Rock he was at a loss to understand the concern at his late arrival. 'You ought to know nothing could happen to me!' he expostulated. 'I've just been working my way out!'

1. Staff members brought from Perth were C. P. (Con) Mathea, paying and receiving officer (later Auditor General), G. C. Stokes and E. A. O'Grady. Everett and Loftus continued to assist. Miss Jessica Stewart was appointed as typist and S. C. McIntyre was later recruited as junior clerk.
2. Erected as a school in 1894, with additions added in 1903. Converted from school to court house in 1905.
3. Personal communication.
4. See M. Rostovtzeff, *The Social and Economic History of the Roman Empire,* revised by P. M. Fraser (Oxford, 1957), vol. 1, pp. 198-9.
5. Matt. 6:24.
6. Matt. 22:21.
7. Miss Flute had been employed in the Lands and Surveys Department, Bridgetown, from 1923 and followed her parents to Narrogin on her father's transfer in employment to the Western Australian Government Railways.

11

'Culbina' 1922-1934

Mitchell was approached by his landlord, Guy M. May, in February 1922 and given first offer for the purchase of 'Morack', which for business reasons Guy was forced to place on the market. With the threat of a transfer in his employment still hanging over his head, Mitchell was unable to take advantage of the offer and moved temporarily into the only accommodation offering, the recently acquired caretaker's cottage on the Greater Sports Ground.[1] He was not very happy about the move—although the Greater Sports Ground Board was not yet able to afford the employment of a caretaker groundsman, he felt that as a member of the Board he was taking advantage of his position, and earnestly sought another place in which he might permanently settle his family, at that time consisting of his wife; Joan, now aged fourteen; Nancy, twelve; Peggy, ten; 'Ginger', six; and 'Bricky', three. Nothing offered until early in 1923 Ross East and his family vacated the stone building in Egerton Street owned by the government.[2] The Mitchell family moved in. During the succeeding winter, Ginger and Bricky contracted pneumonia and, rightly or wrongly, Ida blamed the house—it was cold and damp—and she insisted on their finding another place where there was plenty of sun and fresh air.

Mitchell's close friend, Ted Hardie, had the answer. He was the resident executor for his late wife's estate, 'Culbina', comprising Dumberning Lots 95, 142, 163 and 220, of some 400 acres, 3 miles from town on the Clayton Road.[3] 'You'll do us a favour, Charles, if you'll take a lease and look after it', he said. No information is held as to the rental, but Mitchell would have been too independent to accept anything except a reasonable price. It was, in fact, the first break in his efforts to obtain a second income towards providing for his family's schooling. He always had a few cows and calves and now, by purchases from 'Micky' Brown, they were built up to dairy-herd size. Mitchell was an experienced and able dealer, and sheep were gradually added.

To further supplement his income, aided by the fact that he now had more time at home, he accepted the position, long available to him, of handicapper to the Great Southern Race Clubs. His fees were based on the number of horses to be handicapped, and his tremendous memory for performances was invaluable to him. The appointment was popular, although he used to say that every time the handicaps were published, he left town for a week to let the bookies simmer down.

Joan had already made her contribution to easing the family budget by winning a scholarship to Perth Modern School, tenable for three years commencing 1921.[4]

Mitchell would never own to being soft or sentimental, and with his children he covered these traits, of which he certainly was possessed, with his own inimitable and colourful brand of persiflage. He remembered his schoolboy embarrassment over the nicknames evoked by his red hair—until he started school he had been addressed in the family circle religiously as Charles. At school, 'Charles' soon went by the board. He was 'Ginger' Mitchell, which he eventually accepted, but he would never take 'carrots' or 'pink top'—they were fighting words. He was determined to save his sons this early embarrassment; from birth, Philip was called 'Ginger' and Charles was called 'Bricky', and they were never referred to by the family, in the house or outside, in any other way.

It is possible that even today Bricky's very wide circle of friends and acquaintances, while aware that his initials are C.R., haven't the foggiest idea what they stand for. Bricky himself, up to the age of thirteen, was sure he had been christened Charles Bricky and was recorded in the school register as such.

Apart from settling the matter of nicknames, Mitchell contended that he had contributed materially to their early training also by presenting each with a pocket knife on his first birthday, and a tomahawk on his second—which makes the nature of the 'early training' rather open to question.

All members of the family were keenly aware of his predilection for 'redheads' and when Nancy was waiting for her first child, she spent the last few weeks knitting a flaring red jumper in an effort to influence Mother Nature. As it turned out she was wasting her time, except for having a red jumper for the blackhaired baby to wear. Mitchell's wire of congratulations was typical. 'Red or black? Congratulations anyhow.'

It was at Culbina that the family became a team. Ted Hardie's words, 'keep the place in order', were taken literally. Stone and root picking and the eradication of poisonous weeds under father's leadership became a game in which everybody joined; a competition, always carrying a prize, to see who would make the biggest heap, with dad never winning but often caught making surreptitious contributions to someone else's heap. Stooking hay was always done at the weekend, as a picnic to which the children's friends looked forward to being asked; 'Mummy' Mitchell was famous for her biscuits and generosity. Mitchell seldom left the farm over the weekends, as there were ample chores to fill the daylight hours.

It must have been a wrench for Mitchell voluntarily to give up watching the Imperials Football Team play: he was a foundation patron of the Club and knew all the players, in whom he took a personal interest, most of them being sons of his particular friends. However, consolidation of support for his growing family took precedence over almost everything else: indeed, one of the few extravagances he permitted himself was country membership of the West Australian Club, Perth, which he took out in 1929 and maintained till his resignation in 1960.

I was transferred from Bunbury to the Narrogin branch of the Agricultural Bank in June 1931. Mitchell was away from the office the first week of my arrival. I well remember how the staff spoke of him, either as 'C.D.M.' or Mr Mitchell. There was respect in their tones, and one felt the warmth of their regard. Naturally, I wondered: 'What is he like?'

I needed little persuasion to play for the boss's football team. These were the early days of the Depression—young single men were being found redundant within the Public Service. I did so want to stay in Narrogin—here was the first permanent position presenting itself since Group Settlement had been absorbed by the Agricultural Bank on 1 October 1930. The following Saturday Imperials won their first match for two seasons against Highbury, and I confidently thought this would merit a mention from Mr Mitchell on the next Monday morning. I'd heard he returned from the country. But no—such was not the case. The first thing I knew was a big man, with a warm voice, and a firm hand-clasp looming over my table: 'Good morning, Mr Gabbedy, settling in alright?'

'Yes, Mr Mitchell.'

'Good.'

There was no being sent for, no introduction by a third party. Somehow I felt I'd been accepted.

One did not bother him: at my level one respected his many worries. The approach was through the paying and receiving officer or the clerk-in-charge. Mitchell was aloof but not distant; the warmth of his personality pervaded the office, which was not the same when he was away—something was missing.

One morning in January 1932 he came to my table, soon after nine: 'Mr Gabbedy, I want you to go with Mr Kidson from Dalgety's and represent the Bank in some reported wheat stealing at Lake Biddy.' I, who trained in the South-West, hardly knew what a bag of wheat looked like!

Nothing more, nothing less, and away I went. The trip had been anticipated to last one day, but it was the afternoon of the third day before I reported back.

'Hallo, Mr Gabbedy—a good trip?' was all Mitchell asked.

'Yes, Mr Mitchell.'

'Good.'

Nothing more, nothing less. It was my first personal example of his greatness: a minimum of direction, a maximum of trust. Years later, during World War II, I realized that these two ingredients, more than anything else, promoted initiative and loyalty.

There were other sides to him, of course. Mitchell and football combined to become one of the minor bugbears of my first three years in Narrogin. He became the bane of my life. Each Monday morning he happened to be in the office he would proceed to tell me how badly I played, what a poor example I set in my year as captain, and the wrong tactics I employed, even as coach. There seemed no acceptable answer to such a critic until I discovered that over the first three years of my playing he would not have seen more than three matches, but his family were firm followers and he had his friends' opinions as well to draw from. While Mitchell undoubtedly enjoyed his football vicariously, the delight he obtained from pulling his friends' legs was real and personal. His homily invariably concluded with the admonition: 'You'd be lucky to get a game in my day.'

My revered father had a similar method of enjoyment. In our house, while it was acknowledged that football was played and a certain amount of time spent in training, it was never officially recognized to the extent that the subject was ever discussed in Dad's presence; our thoughts and time, in his oft-repeated opinion, should be given to higher and better things, such as our school books. Imagine my dismay and concern, during one very heated match, when I observed my father on the ground offering to fight the father of one of the opposing players, who had climbed over the fence with the stated intention of taking to me for having knocked his son over several times. I was too petrified to go to his help, and was delighted and relieved when the two old fellows were taken back into the crowd. When I told my mother about it afterwards, she advised me never to speak of it to Dad; he was entitled to his dignity. She had suspected he might have been going to the matches, his gruff subterfuge being: 'I think I'll go for a walk this afternoon.'

Some thirty years later, severing a life-long family tradition of support for South Fremantle, I became a member of the executive of East Perth Football Club. Although my father persisted in calling me 'traitor' and 'renegade', we found out quite by accident that our parents spent every Saturday afternoon with their ears glued to the wireless broadcast of the East Perth game, telling their beads as their contribution towards the team's success, and thereby helping their eldest son. Nothing would do but we then must twit Mum and Dad about their having changed their affiliations. Dad's defence was attack. 'Your mother's not much good as a barracker. During the last quarter of each match, when you really need help, she gets excited and forgets to pray.'

'Hmph', Mum replied, 'what about the day of the Grand Final, at half time, when you said you couldn't stand it any longer? East Perth were too far behind to have a chance, and off you went for a walk. When you came home, you asked what were the final scores. When I said East Perth had won, all you said was: "The older you get, the sillier you get, you can't even hear properly." '

Mitchell and my parents had a lot in common, and one thing in particular: neither ever felt that any explanation of appreciation or loyalty was necessary to the one appreciated or loved.

A PERPETUAL CHORE AND MAIN SOURCE of revenue at Culbina was the making and sale of butter, under Mother's supervision, up to forty pounds a week; Culbina home-made butter, in its own special wrap and at 1s 6d per lb, enjoyed a great demand. The children milked before and after school and work; separating was done by hand and the butter was always made in a small wooden churn.

Ginger was showing signs of Mitchell's love and eye for stock. The dairy herd was his particular care and a cow of particularly mixed breed, named 'Patsy', was his favourite—possibly because of her very placid nature. Came Narrogin Show time, and Mitchell, with his usual enthusiasm, was extolling agricultural shows and challenging the children to prepare entries to see if they could win prizes 'like your mother with her butter'.

'I'll enter Patsy', said Ginger, then aged ten.

Mitchell's heart sank, as Patsy was no show piece. Nevertheless, he wouldn't discourage the young hopeful, so having instructed him in preparation of a beast for showing—even including polishing her horns and hoofs—he sent the boy off on his own to the secretary, Jasper Stone, to enter Patsy in the 'Cow any Age' section. Mr Stone did his best to help with the entry form and asked the youthful competitor:

'What is her pedigree?'

'I don't know what that means', said Ginger.

'Who was her father, then?' Mr Stone tried again.

'Oh', said Ginger, 'that's easy. Micky Brown, of course!'

One can imagine what pleasure Mitchell had in recounting this to all his friends, especially to Micky, and with what pride Ginger received the prize his father's friends hastily donated for 'the best got-up cow in the Show'.

The children loved the life except when Mitchell was away. They lived 3 miles from the school, and then had to walk it both ways, chancing the occasional lift that might come along. All was once again right with the world when he came home; they would sit outside the Bank waiting for him to knock off to drive them home, then share their daily chores with him.

The back-stop at Culbina was, of course, Ida, who held together the home and family life, so good and relaxing for Mitchell. The prim young lady he had whisked off to the altar while he was still only a cut above a farm labourer was now the mother of his five growing children, the keeper-of-the-fort while he was away, the sounding board for all his dreams and frustrations—in fact, by some curious dispensation of Providence, the perfect complement to the man and the times in which she lived with him, in endless mutual endeavour for their children and harmony between themselves.

Nevertheless, the years and the work had told on her; she did not enjoy good health, although, of course, she made little of it. Doctor Mackie once said to her: 'You should have been dead three times to my knowledge, but you just wouldn't give in.'

'Oh no, Doctor', she said cheerfully. 'I've just never had time to die.'

She never complained of her husband's long absences, but quietly and consistently extolled his virtues while he was away. To this day his children tell of the quick glances they often took in their early childhood in the hope of seeing Dad's 'halo'; it was quite a shock, in later life, to find that the same Dad had human weaknesses.

Mitchell taught all the children, boys and girls, to ride. 'Keep your hands down and sit up straight', were the two cardinal rules, but the first commandment was, 'look after your horse before yourself'. Heaven help the child who forgot this—Mitchell couldn't, so 'it never happened more than once'.

As the three girls grew older, Mitchell was their despair because of his attitude to clothes, and for another trait in which, paradoxically, the boys took considerable pride. He never took the slightest care of his clothes and never worried how he looked—except for clean collars and clean boots which were both an absolute must. Most fastidious and immaculate in his personal cleanliness, he was just as likely as not to get under his car in a new suit which, he insisted, must always be tailor-made; first by Harrold's of Perth, later by the W.A. Tailoring Coy. A telegram: 'Send me a new suit, same size', was all that was necessary. His hats were always out of shape after the first wearing—possibly from his penchant for pulling them far down and then sideways on his really outsized head.

Boots were his passion: light and well made for general wear, stout and well made for field work, and both always highly polished. Socks had to be wool, and hand knitted—his younger sister, Mrs Nora Rae, always had a pair for him on birthdays and at Christmas. As a result, and because he looked after his feet, he walked literally thousands of miles but did not have either a corn or a callous until after his retirement.

If the girls could do nothing about their father's habits of dress, they could do even less about the second lamentable failing which, they contended, somewhat dimmed the reflected glory in which they bathed, on race-days, as the children of the handicapper and starter. Their relationship with such a puissant official ensured them front seats in the Race Club stand, but enjoyment of the social prestige was marred almost beyond redemption when it came to the start of a race within earshot. There were always several metropolitan jockeys in each race, 'city-slickers' determined to take advantage of their less experienced 'country cousins', and to humour rather than take notice of the country starter. Mitchell, on the other hand, was equally determined to be the boss, and his commands inevitably became louder, more highly coloured and more explicitly descriptive. Ida and the girls squirmed in their seats but the boys glowed with pride; no other Dad could talk like theirs.

IN THE SOUTH-EASTERN CORNER OF CULBINA there was a large dam which Hardie had created by throwing up a wall across a gully with drays, shovels and scoops. When it was filled it covered an area of nearly 4 acres, with 25 feet of water at the wall end. Without travelling 35 miles to Lake Norring at Wagin, the local youngsters then had nowhere else to swim, and they took full advantage of it. Occasionally Mitchell would rave and rant: 'That's the last time I'll let anyone swim in my dam; cutting the fences to drive cars in, frightening the stock and not letting them drink, never asking permission!'—but that was as far as it went. Far from keeping them out of his paddock, he built a spring-board and put in a gate from Clayton Road for their use, and provided a bough shed to shade picnickers. It must have cost him a deal in loss of time and money, but that was not the point. There was nowhere else for the kids to swim and that was what counted with him.[5]

Pat McCormick, born in Narrogin, now wife of the Hon. F. J. S. Wise, M.L.C., has fond youthful recollections of Mrs Mitchell with her 'sweet soft ways', and 'Mitch', who was a 'dear'.

As teenagers, Pat and her friends swam in Mitchell's dam, and it was for their pleasure that Mitchell erected the first springboard. It was made from two mallet saplings, across which Mitchell had nailed slats taken from beer cases. While Mitchell was able to select saplings of equal length and girth, he was unable to obtain two of the same springiness and, after hammering the last nail, he had to show the kids just how good his masterpiece was. As he got to the end, one side dipped and Mitchell fell in with a great splash. Not deterred in the least, he sat on the bank waving his hat around to get the water out of it, and confessed to the kids that he was not much of a diver—and as a matter-of-fact that he was not much of a swimmer either. The kids offered to teach him, and many were the happy hours spent this way, enlivened with Mitchell's infectious enthusiasm. He practised the various swimming strokes under the guidance of his youthful admirers and advisers and, once having mastered the balance of the spring-board, dived, although with far more splash than skill.

Mitchell departed from his fixed rule of never staying at farm houses when his young friend, Emmet Brown, took up the managership of Micky Brown's 19,800-acre Jitarning holdings in 1923. The homestead, 'Woolgring', was 6 miles north of Jitarning siding.[6] The unmarried Emmet welcomed Mitchell's company and, besides, Mitchell could always borrow a saddle hack from him. He called once when Emmet and the shearers were crutching at an outcamp, and accepted the invitation to stop off and have a cup of tea. It was near 'smoke-oh' time, so Emmet left to put on the kerosene tin of water to boil, returning afterwards to help in the shed. A few minutes later there was an explosion, which completely destroyed the open fire-place and lean-to. Emmet had put on a quarter-full tin of petrol in mistake for water. Once Mitchell had ascertained that no-one had been hurt, and that damage was only minor, he began to relish the incident. It was right up his alley for embellishment in the telling for his friends' delight and Emmet's embarrassment, and he told it with increasing zest and decoration—until the tables were turned on him on the occasion of the marriage of Emmet's sister, Mabel, to J. L. (Dip) de Thierry in 1929.

After the reception, and following the usual custom, the wives were driven to their respective homes while the husbands adjourned to the club, and when that closed, to the home of another Brown brother, Paddy, whose wife was away. Several of the party elected to sleep the night at Paddy's, and knowing that nobody would leave for home while there was a drop of beer to be had, Emmet hid half a dozen bottles in what he considered an extremely novel and secure place—they would serve as a 'hair of the dog' in the morning.

Although Mitchell did not leave Paddy Brown's place until 3 a.m. he was up again at dawn—as was his wont—and his first thought was for his friends. 'Poor beggars, they'll feel dreadful when they wake up, and no-one to look after them', he said to Ida, himself secure in the warmth and satisfaction of his own smooth-running, wife-oiled household. 'I'll just slip in and grill them some steak for their breakfast.' He picked up the steak en route to Paddy's from the rear of 'Butcher' Brown's shop, and when he had lit the fire and made a pot of tea, he woke the sleepers. As one of them recalled it, he had just walked into the bedroom and said: 'Here's something that'll do you good. Drink it up and I'll have some juicy grilled steak for you', when there was a terrific explosion in the kitchen. The five of them raced out, to find the oven door blown off its hinges: Emmet had hidden the beer in the oven. Mitchell never again referred to either incident.

He was a gregarious, fun-loving man, an asset at any party or dinner, the leader in the jocular baiting of bridegrooms, guests-of-honour, trophy winners and all poor souls who ever had to stand up and somehow fill at least five agonizing minutes beginning: 'Mr Chairman, Ladies and Gentlemen . . .'

He was a great believer in giving credit where it was due and despite an over-full schedule, he made a special effort always to attend 'Farewell' and 'Welcome' functions, and he sponsored and contributed to countless public acknowledgements of outstanding behaviour by fellow townspeople. For himself he shunned publicity and until 1926 was successfully reticent. His emergence from this self-imposed obscurity is remembered as one of the outstanding civic occasions in Narrogin's history.

Ted Hardie, his landlord at Culbina, was influenced by his wife's declining health to move to Perth, and as Mitchell had subjected him to a traditional roasting at a farewell dinner tendered him by the members of the race and polo clubs at the Duke of York Hotel on April 1, he was looking for a way to pay his friend back in his own way.

His opportunity came a couple of months later when, after having settled in Perth, he decided to take his family on an extended visit to the United Kingdom, and used the occasion to inveigle Mitchell to another farewell dinner—this time a function in his estimation long overdue, and one in which he and his friends might pay a tribute to 'Mitch' of Narrogin, the man who had done so much for—and to!—his associates, but who so far had evaded thanks and escaped retribution.

The invitation dinner to a 'red man' who was also 'white' took place in the Duke of York Hotel on the evening of Thursday 27 May 1926, the dining-room being filled by more than a hundred leading residents of the town and district.

The toast list and menus were printed in red, which colour predominated in the table decorations and decor. It is certain that Mitchell did not know that he was to be the guest-of-honour, and it would be of great interest to know just how he was inveigled into attending. He must have been kidded with some specious excuse: he never would have been persuaded.

The *Narrogin Observer* (4 June 1926) recorded that: 'It is safe to say that the gentleman so honoured—Mr. C. D. Mitchell, to wit—was the most uncomfortable one present, and those who know of his retiring, modest nature will readily appreciate the difficulty in getting him to attend the function.'

Great was the praise and many the laudatory remarks showered on Mitchell by the various speakers. Naturally they contained a good deal of ribbing and facetiousness—all of his friends had been on the receiving end of his wit and fun-making at some time or another, and here was the ideal opportunity to even the score. His favourite public reiteration that 'no-one could have too many children', and that he, personally, 'would have liked to have had a hundred' was the favourite theme in the all-male gathering, and it was given great play.

One speaker remarked that he was disappointed that evening that there were not more redheaded men present to share in the agony of responding to the main toast, but he felt that was no fault of Charles Mitchell. Anyway, he continued, there was something in being redheaded. Any man could have black hair. Christopher Colombus was redheaded, and after he discovered America subsequent visitors found the country to be populated entirely by red men.

'That should make you jealous, Charlie!' the speaked concluded.

A presentation on behalf of the Narrogin Race Club, Trotting Club, Polo Club and Agricultural Society was subsequently made by E. B. Johnston, M.L.A., who warmly commended Mitchell's services to these bodies and to every public movement for the welfare of Narrogin. The presentation consisted of a very handsome case of pipes and the latest in stop-watches—one that would 'time every horse engaged in the race'.

The years roll by and ranks are thinned of those who attended Mitchell's presentation dinner at the old Duke of York, but those who survive recall it, to a man, as a remarkable night in which the wit, humour and sincerity of the often extempore speeches mirrored the remarkably high level of natural social deftness which was so often, in those times, a feature of small country centres. To the end of his days the night was one of Mitchell's fondest recollections.

THE FAMILY AT CULBINA WAS GROWING UP, and in the manner of families the whole world over, was scattering to school and work. Nancy, enrolled at Albany High School in 1927, completed her Leaving Certificate by the end of 1928 and took up school teaching. Her first appointment was as a monitor to Wickepin. After teaching at Northam for several years she was transferred to Tinkurrin, where she met her future husband, H. F. (Hal) Easton, a member of the Fremantle pioneer family.[7]

Peggy was booked in as a boarder at the Church of England School for Girls (now St Hilda's) at Mosman Park, in 1928. She left before the end of March 1929, to take the position of secretary with Mr Shugg, Principal of the Narrogin School of Agriculture. The Shuggs, close friends of the Mitchell's, not only boarded Peggy but also put up her horse, which she rode home to Culbina each weekend.

Ginger and Bricky just had to follow their father to Hale School, 'to learn how to meet people'. Ginger attended in 1931 and 1932 and Bricky followed him in 1933 for a two-year term.

Following her wedding in 1933 to Colin Wishart, Joan had gone to live in Kondinin: and with the boys away at school and Peggy home only on weekends, the farm became a lonely place for Ida. Mitchell looked around for a house in Narrogin township, and early in 1934 purchased, for £440, in Mrs Mitchell's name, the residence of W. E. C. Moss, ex-town clerk and surveyor. On the corner of Earl and Furnival Streets, it was possibly the first brick house built in Narrogin. The two adjoining blocks[8] were bought as a matter of course. The family would have been lost without the company of a couple of their horses and a house cow, but the sheep, whose numbers had now been increased at the expense of the dairy herd, were transferred to a leased property near Boundain siding. The new home was soon humming with the activity and involvement—and trouble—which seemed to gather like star-dust around the Mitchell constellation.

Ginger and Bricky, home for the May holidays in 1934, were told to take the horses and go out and have a look at the sheep. Having played around until it was too late to go on horseback, they borrowed Dad's car, a 1929 Model A Ford with canvas top, and in their haste turned it completely over 9 miles from town.

Although, happily, neither of them was injured, the car was a mess: the hood a total write-off. Somehow they got it back on its wheels, borrowed petrol and oil, and made for home. As their father was at the Club, they put the car in the garage and gratefully retired to bed. The first thing Mitchell noticed, next morning, was the unusual circumstance of the closed garage doors. 'What's wrong with the car?' he demanded, suspiciously putting two and two together.

'What's the trouble Charlie, dear?' enquired Ida.

'That's what I want to know. The garage doors are shut', Mitchell bellowed.

Confessions were forthcoming, but they only added coals to the fires of his wrath. 'Do you fellows think I'm made of money and take me for a joke?' he demanded. 'I'll show you.'

'Please, Charlie', Ida pleaded, 'they didn't do it on purpose.'

That fractured Mitchell. He gave up, but he carried the story into the office—with all the trimmings of what he said to the boys and what he was going to do about it. After fuming and raging, savage old Mitch quietened down. 'Poor little beggars', he mused. 'They must have been terribly frightened out there in the dark, walking to Hallett's for help!'

That evening, following token rumblings, he gave the car to Ginger, who had it converted into a utility. Next day Mitchell bought a new car—for his own exclusive use, he thundered—but of course he never achieved that ambition.

The incident was one of Mitchell's favourite stories—told in his own inimitable style. What terrible sons he had, how they were doing their best to ruin him; but they couldn't trick their 'old man'. He 'knew something was wrong with the car as soon as [he] saw the garage doors were closed, for the very first time'. Inevitably he would conclude with his own version of his lecture and disciplinary action; finishing with: 'I've taught them a lesson they'll never forget.' No doubt he had—a lesson on just how soft the old man was under all those fireworks.

With the birth of daughter 'Topsy' to Joan and Colin in 1934, Charles and Ida became grandparents for the first time. Ida loved the title 'Granny', by which she was known to one and all for the rest of her days. Possibly irked by the baby's black hair, Mitchell firmly declined the title 'Grandad': 'I want the grandchildren to call me "Mitch" as everyone else does', he declared. Nobody recalls who sponsored or suggested the change from 'Daddy' to

'Mitch'. With any other family, such familiarity could have detracted rather than added to a mutual love and deep respect, but all they remember is the naturalness and ease of it.

Both the boys, like their father, preferred the wide open spaces to academic studies. After leaving Hale School, Ginger enrolled in the two-year course at Narrogin School of Agriculture, and after that he went to work for Uncle Tom (Mitchell's youngest brother, who had resigned from the Bank of New South Wales and purchased a farm at Culbin). After some years he left to gain experience with Harry Biggin (stud sheep breeder) of Kondinin. From there he joined the staff of Elder Smith and Co., where he was employed when, after the outbreak of World War II, he enlisted in the Royal Australian Air Force.

Bricky completed his two-year stint at Hale School, mostly under compulsion, during 1933 and 1934. He joined the outside staff of Elder Smith and Co. in the following year. Today Bricky is manager of the Machinery Division of Westralian Farmers' Co-operative Limited, and an unmistakable chip off the old block. Possibly, his mother's recognition of this likeness laid the foundation for a facet of his accomplishments of which even many of his closest friends are unaware. He is—or was—a pianist. His fond mother, seeking an early refining influence, sent him to the Sacred Heart Convent in Narrogin for several terms to study music. This period of his life he had successfully obliterated—the boy pianist in the country town of those days had no easy row to hoe—until recently when quite by chance I engaged, in conversation, a former fellow pupil of his—then Elva Golding, now Mrs A. E. Johnson. I had not known of Mrs Johnson's acquaintance with Bricky, and was even more surprised when she told me it dated from the music room at the Convent in Narrogin.

'Bricky was the only boy I remember in the music class', Mrs Johnson recalled, fondly. 'We were very proud of our Little Lord Fauntleroy, particularly on the occasion when the Trinity College Examiner, Doctor Boyd, came to the town and Bricky was chosen to go down to the Horden Hotel to bring him and his wife up to the Convent for afternoon tea.'

Willy-nilly, I felt that this charming little vignette of Bricky in his best outfit trotting down the dusty road toward the hotel (no good Catholic convent girl would ever be sent into the environs of *that* Sodom, for whatever purpose!) should be preserved and presented for the delectation of mates who up till now knew nothing of this refinement in his upbringing.

The halcyon life of growing family and community interest at Culbina was approaching its end in the darkest days of the Depression, and the ordeal of the Royal Commission on the affairs of the Agricultural Bank was under way.

The Commission was a watershed in the life of development and progress which had been Mitchell's to this date: he met it and survived it, but it changed the course of his life's action and direction, and before discussing it I feel it apposite to develop further some of the characteristics and connections of the man on whom it had so profound an effect.

This I have done in the following two chapters. 'The Two Brothers' juxtaposes Mitchell and his more famous brother, James; and 'Horses and Polo' shows Mitchell in association with one of the most abiding loves of his life: horses.

1. Situated on Narrogin Town Lot 496, it was purchased from the estate of William Martin Judd on 17 October 1921 for £450. The cottage is in Clayton Road, adjoining the existing football and trotting ground.
2. Narrogin Town Lot 33. It later became the District Office of the Lands and Surveys Department, and today is occupied by the Public Works Department, Architectural Division.
3. Ted and his brother-in-law Henry Ferguson Denny were executors for Ethel Constance Hardie, the land being left in trust for the children. Hardie purchased Lot 142 in 1907 and built a four-roomed stone house for his bride. The property was named 'Culbina', the native name for 'rocky hill' earlier given to a paddock on their Boodarie Station, and is now held by Byron E. Hardie, son of Ted Hardie.
4. Joan passed her Junior Certificate examination in 1922 and after one more year's schooling (to complete the scholarship) took up employment with Elder Smith and Co., Narrogin. There she met and married Colin W. D. Wishart on 29 April 1933. There were two children: Janet (Topsy), born 4 October 1934 and Michael, born 4 December 1937.
5. Mr and Mrs George Harry Hewett had provided a swimming pool in the yard of their home, on Narrogin Lot 730, at the top end of Gordon Street. The pool, dug by hand, was fed from a nearby well; the overflow went into the railway drain.
6. Michael Brown discovered 'Geetarning' Soak soon after his arrival in Narrogin and following the usual practice, promptly selected the surrounding area. The homestead was built on the high ground to

the north. 'Jitarning' became the modern day spelling of the soak's name.

7. They were married in Narrogin on 16 December 1937. There were two children of the marriage: Ian Mitchell, born 30 March 1939 and Judith Marian, born 6 June 1941.
8. The three blocks were Lots 85, 86 and 87, with the house on Lot 87.

12

The Two Brothers

Charles Mitchell's life was made no easier by the fact that his elder brother was a Minister of the Crown, especially as both were hard-headed, blunt, honest, pragmatical and vigorous men. However, because Charles liked to work alone and shunned social affairs, the two brothers met but seldom and personality clashes were rare.

On 17 May 1919, and in his fifty-fourth year, James became Premier and Treasurer for the first time, retaining the Lands and Repatriation portfolios. In that year also he received the honour C.M.G. His government was returned on 12 March 1921, and shortly afterwards he was knighted and created K.C.M.G.

At the outset of his second term he was determined to put into practice, immediately, at least one of the recommendations of the Royal Commission on the Agricultural Industries of Western Australia in their reports of 1917 and 1918.[1] This referred specifically to the necessity to institute effective liaison and an improved advisory service which would promote a closer understanding of the farmers' problems. The answer, unquestionably, lay in the decentralization of the Agricultural Bank, through the establishment of branch offices at selected country centres.

Charles Mitchell had thirteen years of service behind him under the Agricultural Bank Act, which had governed employment of all field staff, and apparently he was considered an integral part of this plan. On 1 July 1921, at the age of forty-five, he was appointed to the permanent staff of the State Civil Service. At the same time he was confirmed in the position of district inspector at a salary of £300 per annum and was provided with a government car.

It would be fairly natural to conclude that Charles owed part or whole of his preferments to the influence of his eldest brother, but never was any assumption further from the truth. Sir James was accused of many things during his political career, but never of nepotism. It could not have been that he was afraid of criticism. If so, he certainly never at any time gave his political opponents that impression. If anything, he appeared opposed to his brother's preferment and Charles once expressed this conviction to his family. Not only did thirteen years divide the brothers in age; they were poles apart in temperament and inclination. Charles hardly knew James, who had left Bunbury before Charles was nine years old. Whenever they met in later life the elder's attitude was that of austere patriarchy, so different from that of the eldest sister, Kate: Charles was in his eightieth year when he last saw Kate (then in her ninety-second year), calling in en route to the Bunbury Races. 'Goodbye, Charles', she said. 'Have a good time and behave yourself. Oh! Charles dear, have you got a clean handkerchief?'

Charles described his brother as a 'hard man' and told, perhaps with disdain, of their first meeting after his appointment to the permanent staff. There was no word of congratulations, just: 'Charles, if I ever catch you driving the government car on private business or for pleasure, you are reported.'

Charles held a personal belief that merely to possess red hair bestowed certain superior qualities on the possessors. He knew nothing of the science of genetics nor did he care. This conviction, and its very probable reiteration, may have caused friction and antipathy. Of the thirteen Mitchell children who survived infancy, seven had black hair and six had red. There were two distinct types in the family, the personality, character and traits and even the appearance of the one group being almost the antithesis of the other. As Charles would relate: 'Jim led the "blacks" and I led the "reds", and six reds were always too good for seven blacks.'

R. G. (Reg) Courtenay, recalls that when he was chief accountant of the Agricultural Bank in the early 1930s he kept an appointment with the Premier, Sir James. When he arrived, Sir James's secretary said: 'I'm afraid you'll have to wait, Mr Courtenay. His brother Charles has just gone in unannounced. Have a peep around the door, I'm sure you'll be interested. I've just served them tea.' Courtenay peeped.

Charles had recommended dispossession of a farmer in his district. The trustees accepted the recommendation but later, at the request of the Premier, reversed their decision. Charles was not accepting 'this type of interference' lying down, so he had come to Perth and straight to the source of the intervention.

There sat Sir James shouting: 'He stays!' and emphasizing the point with his fist on the table. Charles, just as loud and emphatic, was shouting: 'He goes!' and thumping the table even harder. The tea, undrunk, jumped from the cup into the saucer and from there to the table. The reds won, and the man went.

A final instance perhaps emphasizes the width of the breach between the natures of these two outstanding men. In 1941 Charles took a young friend, for whom he was seeking a reference, to Government House. An excellent reference resulted; Sir James knew the young man and his family very well. As they came out Charles turned and said: 'Did you listen to him? Talks like a Year Book! Drives you mad—always been the same.'

On the one hand was the elder brother, imperious, gifted with clear and broad vision, confident, commanding, egotistical and self-seeking perhaps, but a born leader with so much to do and so little time to do it in. On the other was the younger brother, knowledgeable in his own field, a humanist experienced in the fine details of applying simple principles, a natural leader, combining simplicity with militancy, yet so humble, responsible and loyal to his superiors as to seem almost self-abasing.

Sir James was one of the three men I have met during my lifetime who undoubtedly could behold the 'vision splendid'—the ultimate in the inspired plan of each, concerned in the sincere belief that it was for the good of the state and/or the people. The other two are still active in public service and will not be made aware of my opinion until they receive a copy of these observations. A clear mental picture of their goals gave—and gives—them a courage and an all-seeing authority brooking no petty opposition and engendering singleness of purpose above all other considerations.

Had it been possible for these brothers to work in tandem or, better still, in double harness, in the exchange of knowledge and practices, there is no knowing how much earlier our agro-economic history might have felt the changes which have since overtaken it. As it was, the isolation of final authority, entrenched in the centralized administration of Head Office, coupled with the seemingly unbridgeable chasm in understanding between Head Office and country administration, produced an endemic inability to recognize what was happening until it was long overdue.

However, Charles and James, widely different in many respects, had certain things in common—a capacity to 'walk with kings and never lose the common touch', the enjoyment of hunting, a love of the land and confidence in it, and the highest regard for family life—the backbone of the state for both of them.

To each brother farming was a way of life, sustained and perpetuated through family succession. James's conviction that Western Australia's future lay in agriculture and therefrom in the sacredness of family life, was genuine. Although at times it was adapted to political expediency, the circumstances never overrode his motivations. His view was wide, covering the whole of the South-West Land Division. Charles's was personal, at individual farmer and farm level.

Their love of children was instinctive, and was readily realized by the children themselves. They were the Pied Pipers of the Mitchell family.

Sir James's repeated advice to young people was to marry early and rise early—suggesting 5.30 a.m. as the ideal hour for rising. His continuous exhortation was 'populate or perish'.

Although neither Charles nor James actively courted anybody's good opinion, people loved them, and quite often for no very apparent reason. One instance of Sir James's effect on people—sometimes even on people who knew him only distantly—concerns a lady whose early life in Western Australia should have left her with no love for him and the things he represented. She told me about it some years ago. As a girl of seven, she had come out with her family as migrants from England in 1924, to be absorbed into the Group Settlement Scheme established in the Denmark area. The ship berthed at Albany and that evening the family were on their block. It was still entirely in a virgin state and for some weeks they were obliged to live under sheets of iron leaned against trees while a space was cleared and the house erected. After some years of hopeless struggle the family walked off, and the memories must have lingered and rankled. The girl eventually married, and in the post-war years gave birth to her sixth child, a boy. As she lay in hospital in came her 'lord and master', looking, she said, 'as proud as if he had had the baby himself!'

'How's little Donald?' he asked.

'Donald nothing', she said, 'It's James.'

'Where-in-the-hell did you get James from?'

'He's named after Sir James Mitchell', she said, 'the only man I ever knew of who had proper regard for mothers with large families!'

And James the baby remained.

Culbina homestead
A recent photograph of the now unoccupied house

The dedication of Charles and James to their respective professions gave no count to the hours involved. The state Under Treasurer, Edwin Alexander Black, was appalled on coming into his office at the usual 9 a.m. on 18 May 1919 to find the newly appointed Treasurer fully installed, coat off and sleeves rolled up at a table already littered with papers.

'Oh, Sir James', he protested, 'If I'd known you were coming in early, I'd have made a point of being here to give you an idea of what is going on.'

'Mr Black, there's a lot to be done and I've made a start', Sir James declared. 'From now on you won't find any clocks here, and you won't develop corns except on your hands.'

Charles could never have brought himself to talk to his staff so abruptly. His appeal would have been by example, but it would have yielded equal results.

As to Sir James's imperiousness, Sir Ross McLarty, when Premier, recalled a fledgling member's complaining that he did not know enough of what was going on.

'It's just as well, young fellow, that you were not a Minister under Sir James; then you'd have had something to complain about', Sir Ross replied. 'Sir James convened Cabinet meetings at most infrequent intervals and then they were usually on the "volunteer system" ascribed to the Army—you, you, you and you.'

Following his opening the Narrogin Show in 1936, Sir James stayed on the following day to attend a Field Day organized by the Agricultural Society to demonstrate the successful establishment of subterranean clover, in its third year, on Aubrey Hardie's 'Rosedale' property. Hardie's experiments fully justified his confidence, and marked a new era in the development of the Narrogin district.

Sir James, speaking at the conclusion of the day, expressed his pleasure at what he had seen, and went on to state that it was no more than an example of what he had been preaching all his official life. To prove this he made a bet with W. J. (Bill) Corrigan, president of the Agricultural Society: £1 to be paid to the local hospital if a young officer present, who had been trained in Group Settlement, were to prove unable to repeat the injunction that he, as Minister and Premier, had repeatedly passed on to his staff.

I was 'the young officer' present that day, then in my sixth year with the Bank and accompanying the official party as the inspector in charge of that particular section of the Narrogin district. Imagine my consternation when, having quit the official party to be one of the first round the keg, I was summoned back to the dais—the rear of a convenient utility. What on earth did they want?

'Now, young fellow', said Sir James, 'let's see how well you remember your early training. Tell Mr Corrigan what I taught all of you in Group Settlement.'

Thank heaven my memory worked!

'Sir James', I said, 'You used to preach to us at every opportunity that there is no bad land in Western Australia. It was our job to find out what it was good for.'

Sir James, visibly elated, then pontificated: 'Not only should men on the land be not afraid, but they should also live dangerously.' He explained no worthwhile results could be realized unless there was a disposition to take risks. The Lieutenant Governor really loved the land—all of it, swamp, hill, forest, and plain—and would hear no bad word about any acre of it. Charles had an equal devotion. He regarded the wastage or the misuse of farming land as being in the same category as criminal rape.

Although both brothers 'walked with kings'—James more often than Charles—each retained a genuine love of the people. Irascible and demanding as James became, he never forgot a face or a name. His early morning walks along St George's Terrace, with a kindly word and a tilt of his famous boxer hat for everyone, passed into legend. On many occasions, official and unofficial, I was privileged to observe the simple affection of his relations with people of all ranks of life, and to watch how the kindly magic of his personality worked on them. One at least I must relate.

During the winter of 1943, being the only Western Australian officer on board an Australian naval vessel in Fremantle, I was granted three days leave to visit my family at Wilga while the ship was refuelling and storing. Having caught the 11.35 p.m. 'midnight horror' from Perth, I was up at first light next morning, walking up and down Donnybrook station to get warm. Early and cold as the hour was, I noticed that Sir James, with his favourite grandson, 'Dicky' Anketell, was walking in the railway yards while the Railway Commissioner's private coach, which had been provided for his touring convenience, was added to the train.

Later on, somewhere between Katterup and Lowden, the puffing mixed-goods-and-passenger train came to a gradual stop. As passengers poked their sleepy heads out at windows to ascertain the cause, down from his coach climbed Sir James, to be joined by the equally portly guard, whose sole distinguishing item of uniform was his long, black-sleeved cold-weather waistcoat. Together they waddled to the fence bordering the main road, where they were met by an elderly lady and, after an exchange of greetings, given a rather large paper parcel. Back at the train, each climbed into his own compartment and we proceeded on our leisurely way, as though such unscheduled stops were everyday occurrences. In due course, around nine in the morning, the train arrived at Mumballup, and the twelve passengers, including Sir James, walked across to the nearby hotel where tea and scones were served in the kitchen by arrangement with the Western Australian Government Railways. I joined Sir James en route and, acknowledging my salute, he volunteered: 'I suppose you wondered why we stopped back there? I've known that guard for fifty years. He and his wife always have a piece of home-grown pork for me whenever I pass.'

When we reached the hotel the publican's wife was somewhat un-nerved by the presence of a Naval officer, all gold braid and buttons, in her 'country pub' kitchen. She rushed to serve me, hands shaking, tea spilling. Holding her hands with the cup in them, I said gently, 'I think you should give priority to His Excellency the Lieutenant Governor', and then introduced Sir James. That was the last straw. The cup flew a foot into the air and on to the floor. Sir James was quite unperturbed. 'Commander', he said to me, 'you act as my aide, please, and get me a cup of tea.' Turning to the flustered lady he went on: 'Now don't be fussed. Come with me and let me taste one of your tempting hot buttered scones.' In a moment the publican's wife was completely at her ease, chatting with the old man as though she had known him all her life, and no doubt, as they made toward the buttered scones, rehearsing for her friends at the Country Women's Association an account of how she entertained the Governor in her kitchen. He really had a wonderful way with people, either because he genuinely loved them or just because, like the consummate politician he was, he recognized everyone first as a voter and then a person.

Each brother had a fiery temper, not easily roused and not easily soothed. Charles, the complete extrovert, invariably exploded. James, when crossed or frustrated, actually ground his teeth in sheer rage; but there was one time, on a hunting trip, when his righteous rage flashed out—into words and intended action.

Both James and Charles were excellent gun shots, and both enjoyed hunting. Their father, who had brought two valuable shot-guns from England, had carefully trained his sons not only how to handle guns and care for them, but also in the fundamentals of true sportsmanship. James preserved all the ethics. To shoot a sitting duck was a crime most heinous.

Jack Kerr, when manager of Millars' Timber and Trading Company's mill at Palgarup, delighted to tell of the time when he was included in a party of duck shooters, organized by Andrew Muir of Mordalup, to mark a personal visit by the Lieutenant Governor. In honour of their important visitor the organizers had been out some days previously and prepared 'blinds' for each of the party. The shooters had hardly taken their places when a young man to the immediate left of Sir James blazed away at a solitary duck swimming past. James's roar frightened every duck for half a mile around. 'Come out of there, you cowardly skunk', he bellowed. 'Come out and show yourself, and start running so that I can take a shot at you. I'll teach you to shoot a sitting duck!' He was placated only by the party's agreeing to take the young man out of his blind and pack him off home.

James's tongue was not turned against others only, however, and a fine example of his lack of pretension and his earthy nature is provided in an anecdote of R. W. (Wally) Lickfold. Wally, an inspector attached to the Manjimup branch of the Rural and Industries Bank in 1946,[2] received a telephone call from Head Office asking him to place himself at the disposal of Sir James who would arrive at Manjimup on the following day by private rail coach—without any advertisement of his presence—for a couple of nights' duck shooting at Lake Muir.

Wally duly reported for duty the following lunch time, to be greeted by Sir James's general factotum and cook, one Jeffries—an ex-ticket inspector for the W.A.G.R. whom Sir James regarded as indispensable to his comfort. Wally was introduced to Sir James, and the three of them began to pack for the trip: an old packing case in which was loaded bread, steak, salt, butter, tea, sugar and whisky. The camp they set up was as basic as the provisions. 'It was the roughest camp I ever made', Wally recalls. 'Just a tent fly stretched to keep the sun off.' Two Muir boys made up the party. Wally thinks they were from the family farming at the south end of Lake Muir.

It was Wally's duty to prepare the blind in which Sir James might establish himself without frightening off the intended victims, and the old man's instructions were a model of terseness and—when one recalls that Sir James was practically as broad as he was tall—of commonsense also.

'Now look here, Lickfold', he said 'When you're preparing a blind for me, don't get one that I've got to crawl to. You'll be wasting your time. I'm no more conspicuous standing up than I am crawling.'

Charles lacked the theatricalism of his elder brother, around whose name legends have gathered like lichen on a rock. He was, however, just as expert a rifle shot and more practical in his approach. When complimented on his prowess, he would explain: 'Our father taught us well—not only that cartridges were expensive, but that if you missed, you might go hungry!'

Both James and Charles worked long and hard in their public lives—James for more than half a century, Charles for thirty-six years. James became at least a state, and possibly a national, figure, rewarded by his grateful contemporaries with the highest of honours. At the valedictory dinner accorded him by both state Houses of Parliament when he relinquished the Governorship, he said in reply to the many toasts: 'I have worked hard the whole of my life, but never as arduously as I did within the confines of this building.'

James did his work and reaped his rewards and his recognition. Charles worked just as hard, although in a much humbler capacity, but reaped only those rewards generally considered the logical due of virtue: the knowledge, within himself, that he had done what he could, to the top of his bent.

To these rewards he could nevertheless add something else. If the elder brother became history enshrined in Hansard, the younger became a legend enshrined in the hearts of a thousand friends.

It is curious that two such outstanding men should have emerged from that humble, secluded boyhood in Bunbury, and one wonders what unknown forces pushed, or dragged, them both to the top by such widely varying routes. William Bedford, the father, was an unusual man—honest, vigorous, hardworking, intelligent—and Caroline was the archetypal mother, reflecting all the qualities of her men with the added lustre of a bottomless love and patience not only for her own husband and children, but for almost everybody with whom she came in contact. The whole large, busy, united family reflected in turn the forces which went into their birth and upbringing, but none so strongly as James and Charles, and no doubt it was natural that each should have assumed the leadership of his faction within the family: the reds and the blacks—Charles and James. Just as curious as their emergence from their background is the fact that these two men, almost diametrically opposed in every way, should have striven in his own way to achieve so much for his home state.

1. *The Report of the Royal Commission on Agricultural Industries of Western Australia* was issued in two parts, dated 22 November 1917 and 17 May 1918 respectively.
2. The Agricultural Bank changed its activities to a general trading bank and its name to the Rural and Industries Bank on 1 October 1945.

13

Horses and Polo

A honeypot had no more attraction for flies than horses in action had for Mitchell, and it was an affinity which was to bring him not only a great deal of enjoyment, but was also responsible for a considerable degree of his development and progression.

The Narrogin Polo Club was established in 1912, with J. B. Lewis as its inaugural president and A. E. (Ted) Hardie as his secretary—an office Hardie was to fill perennially along with that of chairman of committees of the Race Club.

The sport had received a tremendous fillip in Narrogin by the arrival of the Hardie brothers—Tom (T.A.) in 1907 and Ted in 1909. Sheep-station owners from the north—Warralong and Boodarie, in the hinterland of Port Hedland—they were skilled horsemen and had brought a number of excellent polo ponies from their northern properties. From Messrs Clayton and Rintoul of Narrogin they bought 10,000 acres which they named 'Rosedale'[1] and forthwith plunged into the life of the district, Tom as president of the Agricultural Society from 1911 to 1919, and Ted as chief influence (greatly helped by Guy May)[2] in the revival of the Narrogin Race Club in 1910.

The first polo team—Tom and Ted Hardie, Dr Lewis and the dentist A. C. N. Olden—practised every weekend in a ground they had levelled in Micky Brown's paddock, now the southern end of the golf links and part of the race course enclosure.

Busy as Mitchell was, he literally manufactured time to go and watch them. There he met up with Henry Lawrence Chipper, a High School old boy eight years his junior—but the 'old school tie' was not the link. Chipper, like Mitchell, was a born horseman with a like sense of humour, and penchant for practical jokes—and it was natural that they should become the closest of friends. It wasn't long before they were asked if they would like a game, and not much longer before their blend of youthful skill and superb and dashing horsemanship got them a grading far ahead of the doctor and the dentist.

This contact with the Hardie brothers—comparatively affluent and genuinely altruistic in their community-minded outlook—was one of the important factors of Mitchell's life. Although Tom was sixteen, and Ted four years older than him, they were perhaps the first among all his associates to recognize his strength of character, his humanity and latent leadership, the tremendous power of his selfless example, while at the same time appreciating his shyness, his pride and a humility bordering on self-abasement.

Mitchell's only mount, the faithful, aging Nemo (Nominus displayed no facility toward polo), could not match the pace of the Hardies' trained polo ponies, and Ted, explaining that it would give the No. 1 team more drive and balance, persuaded him to accept the loan of a gaunt, mean looking, blood line, mousey bay gelding of sixteen hands, called Hard Times because it had proved impossible to fatten him sufficiently to prevent his ribs from showing. Mitchell's agreement to stable Hard Times, so that he could ride it for training, provided the Hardies with an excuse to deliver a few bags of chaff, every now and then, as a return for his taking care of the rangy gelding.

A horse of such size was normally not acceptable in competitive polo, and it was a compliment to Mitchell's riding control and sportsmanship that an exception was made. He rode like the wind, devoid of fear, and it must have been an awe-inspiring sight to witness him bearing down, towering over his opponent and riding loosely with the long rein of his old cavalry bridle, the only one he possessed. If there were ever any qualms, only his opponents would have felt them.

Olden was a correct rather than a natural rider and found it difficult to handle his mount—a well-bred chestnut mare of 14.2 hands called Nellie. Ted Hardie bought the mare from Olden and asked Mitchell, as a further favour, if he would accept her as his second string.

With these two horses at his disposal Mitchell became a star, but largely through the tact, thoughtfulness and unassuming generosity of the Hardie brothers—rare qualities which often passed unnoticed but which indicated a nobility of spirit seldom witnessed and less often recorded.

By the beginning of 1912 Narrogin's polo players had become tired of competing among themselves and looked around for other worlds to conquer. Katanning's polo team was then widely regarded as the 'tops' and 'very, very, snooty'; so off to Katanning went Narrogin's team of the two Hardies, Mitchell and Chipper. They travelled the 65 miles in a buggy and four-in-hand of polo ponies, trailing after them a further four polo ponies which, on the way down, they took the opportunity to break into harness. They carried two water-bags: one, unadulterated, for the first four chukkers and the other, strengthened with a bottle of whisky, reserved for the last two. Mighty were the deeds wrought by those doughty warriors in the final chukkers. Three days later they returned home, Katanning no longer the tops and much easier to get on with.

Katanning nevertheless resented their defeat; it rankled. Their resolve to avenge the humiliation was strengthened by the recent settlement in nearby Broomehill district of the Richardson brothers, Edgar and Tom, and George Hardie, brother of Tom and Ted, all three with a knowledge and love of polo from competition in Port Hedland, and a yearning to play again.

The Katanning team travelled to Narrogin the following year, bringing their horses by train, and Narrogin suffered a defeat. Harry Chipper was not available and A. E. (Cock) Robinson took his place. Mitchell often recalled 'Cock's' placid approach to the game, as he rode onto the ground smoking his pipe: 'We got him to put it away, and it must have upset him', Mitchell would explain. 'He only got one hit for the match, and that was when he hit me under the chin and laid me out.'

On 15 March 1913—a Saturday—the Narrogin team went to Northam to play the newly founded Northam team, which had had its official club opening only on January 22 of that year. The match was played with great spirit on the Island farm ground, but it had a tragic ending. John Gollan, of the Northam team, was first rolled on by his horse in the second chukker, and then later received a blow from a stick. He continued to play, but fell from his horse, dead, only three minutes before the end of the match. It is not known whether Mitchell had been involved in either accident, but he stayed on in Northam for the funeral on the following Tuesday, and acted as one of the pall bearers. He never again played polo outside Narrogin, and indeed the accident threw over Western Australian polo a shadow which, with the intervention of the war years, lingered for a decade.

In the matches at home, however, Mitchell continued to exhibit his characteristic fire. Early in 1914, a composite team from Katanning/Broomehill made the trip, again by train. The match is remembered today only as a 'no decision' bout, between Mitchell and Tom Richardson, an opposition redhead. Tom was a thrusting left-hander who hated giving way, as the rules demanded, to the conventional opponent, and he clashed repeatedly with Mitchell. Mitchell, equally thrusting and fiery of nature, soon took exception and made no bones about what he thought of Richardson, his lack of manners and his antecedents. Richardson in turn took exception to the repeated use of singularly descriptive and challenging terms by Mitchell, and the next thing they piled off their horses in the middle of the ground and, standing toe-to-toe, began to punch. The other players, to the dismay of the spectators, soon separated them, and it can be recorded that after the event they 'drowned' the hatchet in lieu of burying it, thereby cementing a mutual and life-long friendship.

Following the outbreak of the First World War polo was perforce abandoned; the district's young horsemen not only rushed to join the 10th Light Horse, but took their ponies with them.

Nellie could not be left completely idle and eventually was put to Micky Brown's Sesame.[3] A handsome chestnut colt was born on 1 October 1918, and named Damascus, in honour of Nellie's first owner.[4]

One game only is recorded as having been played on the old ground, following the Armistice, just before it was ripped up to become part of the new sports ground. The arrangement for the match was probably inspired by the visit of Colonel Olden, who came to dispose of his business and remaining polo pony, Alfred. The Narroginites were very proud of their old associate who, from peeling spuds on his first day's service, had risen to command his regiment.

The 'old fellows'—Tom and Ted Hardie, Mitchell and Olden—played the 'youngsters'—Aubrey Hardie, Paddy and Jack Brown and Jack Stevens. The losers were to pay for a dinner at the Hordern Hotel that night, Frank Johns, the umpire, being included in the party. The 'old fellows' won. No-one can remember the score, but all recall vividly Mitchell's surreptitiously inciting Alfred to buck Olden off. Having got rid of his rider, Alfred took to the bush with the rest of the team in full cry. It was half an hour later before the game was re-started.

The dinner is remembered by the 'youngsters' for the mental anguish they went through as lavish course after course was served, each with its appropriate wine. They had thought 'ten bob each' sufficient and that was about all they had with them. Now they had to draw 'mine host' aside, confess and ask for credit, rather ponderously conceded by the latter who was in the joke. After they had been kept on tenterhooks long enough, Ted Hardie confessed he had not only ordered the special dinner beforehand, but also paid for it.

Polo players never pass away and the Narrogin breed was no exception. There were ponies to be trained and exercised, and spare Sunday afternoons in the late summers would see Mitchell and others out at Jack Higham's property, 'Minigin'. There were good young horsemen coming along who had graduated from horse-holders to trial players and were showing considerable promise. Amongst them were Clarrie O'Connor, Dave Barron, Frank Robinson and the Bradford boys, Stan and Sam, as well as the youthful veterans J. T. (Jack) and H. P. (Paddy) Brown (Micky's eldest sons) and Aubrey Hardie (Tom's eldest son). Mitchell played when he could find time. Hard Times had passed on and Nellie was pensioned off. Damascus had grown into a magnificent animal—15.2 hands—and the delight of Mitchell's horse-loving nature. 'This horse is the best polo pony I ever rode', he averred. The youngsters loved his company, admiring his horsemanship and his devil-may-care riding, and enjoying the stories and reminiscences told as only he could tell them.

However, though polo phased out as a dominating interest, the Agricultural Shows took up the slack for Mitchell. He made a fine picture on well-groomed, gleaming Damascus, and they were central figures when he acted as a marshall and the judge of 'Horses in Action' at succeeding shows. A contemporary described him as one of the two 'prettiest' male riders he had ever observed —possessing the strength and fire of the man and the gracefulness and carriage of the superb lady rider. There was no limit to Mitchell's energy. As soon as he could be spared from his official duties he threw himself into the equestrian novelty events where, mounted on Jack Brown's grey horse Commodore, of 14 hands, he was unbeatable.

Public exhibition and horsemanship were not his sole contributions, however. Elected to the committee of the Narrogin Agricultural Society on 21 March 1914, he maintained an active executive association until 1935. He was appointed to the horses sub-committee in 1919, and became treasurer two years later. It is recorded also that he was one of the driving forces in the re-organization of the Society following World War I. In all, the life membership conferred on him at the Annual Meeting of 19 December 1926 was fully earned and well merited.

Punctilious himself in the matter of attending Committee meetings, and almost always one of the first to turn up at busy-bees to improve the grounds or to prepare them for specific occasions, he had the priceless ability to fire others with his own enthusiasm. Few could deny him a favour, perhaps because it was never asked for himself. His bank staff traditionally manned the ticket offices at shows and race meetings, but it was not a case of 'doing what the boss ordered'. Their loyalty to Mitchell was such that they would have been deeply hurt if they had not been asked.

The volunteer Fire Brigade, established in 1926, was in no way under his direction, but could always be cajoled into staging a pumping practice on the morning of a race meeting, and so laid the dust in the straight before going home for breakfast. Mitchell was always there with them, however, to assist in handling and rolling up the wet hoses—and, from his own pocket, to provide a gratuity for the Brigade's funds.

The levelling of the new polo ground, within the trotting arena, began in 1924. It was played on and further improved the following year and by January 1926 was in perfect order for the initial polo tournament.

Three teams competed, from Narrogin, Broomehill and Perth; Frank Johns acted as umpire, giving satisfaction to all teams. Mitchell captained the Narrogin team (green colours) of Aubrey Hardie, Jack Brown and Dave Barron and, as skipper, his one stentorian direction, given to the team-mate in front, was: 'Ride —ride like hell.'

It was virtually a recreation of Banjo Paterson's famous ballad, 'The Geebung Polo Club'. The Narrogin team sported riding boots and trousers of varying shades and types—but there uniformity ended. Aubrey, Jack and Dave wore unmatched regulation helmets, but Mitchell eschewed such affectation, preferring his inevitable old felt hat, pulled down all round. He also scorned the special short reins[5]—perhaps because of their cost and limited use—and rode with his cavalry reins. He still used the sticks he had had in 1912, having regularly replaced the worn mallet heads with hand-fashioned pieces of york gum.

The Perth team—Norman, Lefroy, D'Arcy and Pearce, all well known in state polo circles—arrived from the metropolis correctly attired in matching riding pants, boots and helmets, their thirteen ponies all expensively rugged and attended by grooms.

Naturally the appearance of Mitchell's apparently hayseed team among these paragons provoked some mirth and good-natured comment, but by the end of the week mirth and comment had changed to awe. The Narrogin team won the three-cornered championship by defeating Perth seven goals to nil on the first match, on Monday, and downing Broomehill by four goals to one on the following Friday. Mitchell himself scored three goals and 'starred in defence', but his glory was shared by the faithful Damascus: when Mitchell's second horse—Aubrey Hardie's Kymac—broke its fetlock in the second chukker of the first match, he brought Damascus in and used him for the remainder of the tournament.

Naturally, the Narroginites were enchanted by their team's victory over the 'snobs' from the city, but in defence of the snobs it must be recorded that certain of the hayseeds' tactics and equipment must have proved somewhat off-putting to orthodox metropolitan players. One item of equipment which made them raise their eyebrows was the whip carried by each member of the Narrogin team—a stick with a piece of rubber hose threaded over the end. 'It wasn't the thundering hooves of the overtaking horse that put you off', said one of the visitors, recording his views of the match, 'It was that damn thump, thump, thump, of those hose whips on their horses' ribs!'

The Narrogin team travelled to Perth the following April to compete for the Gooch Cup, but Mitchell, considering the expense beyond him, stood down in favour of Aubrey Hardie: he also lent Damascus to Frank Robinson for the occasion. The team was defeated in the final by their handicap of two goals.

Nobody who ever had the privilege of riding Damascus ever forgot it. Stan Bradford had the honour for the 1927 and 1928 tournaments, and recalls Damascus 'looking at me out of one eye as though sizing me up—hoping I could at least ride and not pretend to know more about polo than he did. Like his owner, Damascus had an individuality all his own. He was a brainy horse —you know only a brainy horse ever becomes a one-man horse. Once I learned and accepted his idea of playing polo, he was the greatest horse I ever rode. He never took his eye off the ball, and he had tremendous anticipation.'

One night towards the end of 1928 Damascus reputedly broke out of his yard, and was never seen again; many thought he had been stolen. Mitchell felt he could never break in another horse, and gradually lost interest. He had played the game for a long time and, characteristically, to the top of his bent. Although the handicapping system, as known today, was introduced at about the time he retired from the game, he had never played in the metropolitan area and so never had a handicap recorded. His contemporaries, however, concede the rating of a present-day 'four mark'.

Far too many men, when age or other considerations bar them from a hitherto well-loved sport, lose all interest in it, but that was not Mitchell's style. He never lost his love of polo, and maintained it on every possible occasion by frequenting a group of young and enthusiastic horsemen—the McDougall brothers (Ross and Roy), Charlie Hill, Richie and Laurie Wiese, Les Ashworth, Vince Dawson and Roy Blight. These aspiring players, riding borrowed mounts or their fathers' horses, were members of the local troop of the Light Horse and were anxious to improve their riding skills through polo. Mitchell, with his pipe, his criticism and his dry humour, was always welcome. His company and his advice were sought and, like his approbation, were treasured.

With the advent of the Depression competitive polo was abandoned—it became too costly—but it was replaced by the 'Mounted Harriers' organized by Colin (Bluey) Campbell and J. T. (Jack) Brown. Every second Sunday all horsemen were invited to attend a paper chase starting at 1.30 p.m. The course varied from meeting to meeting, with distances from 3 to 5 miles, and ended with afternoon tea. Although everybody knew where afternoon tea would be served, everyone was expected to do his best to find and follow the trail left by the hares.

Les Ashworth[6] vividly recalls the first time Mitchell rode with the Harriers:

> About half a mile from the afternoon tea site Mitch yelled out: 'Come on! Full belt for the last leg! Follow the leader', and off he went, hell for leather, jumping everything in his path. I was riding a horse called Nooka—a good nag, but the trouble was that neither he nor I had ever tried steeple chasing before. You couldn't *not* follow Mitch, though, and I took off after him, just closing my eyes and giving Nooka his head every time we looked like becoming air-borne. I suppose I was lucky to live to tell the tale—but it was the sort of thing that Mitchell would do and get away with. Just being in his company made life a little more exciting.

The Depression killed the carefree Sunday afternoon sport. With farm incomes dropping farther and farther, and the future looking glum, farmers were perhaps too dispirited to make the effort—or perhaps even that trifling expense was too much for men at bed-rock. Whatever the cause, the Harriers disbanded at the end of 1931.

Mitchell became a member of the Narrogin Race Club Committee in 1910 and remained on the executive until his resignation in 1935, by which time he had attained to the position of president.

Lack of time on his part, and the availability of plenty of racing-minded men of executive ability in Narrogin, kept his official activities directed mainly towards the handicapping and starting of the Picnic Race Meeting, held on Easter Monday 1914. 'The meeting was a splendid success', said the *Narrogin Observer* of that time. 'As starter Mr. Mitchell was all that could be desired, sending the fields away in splendid order', and so he continued to be reported in the local newspapers during the succeeding years.

Labor Day (Wickepin, 1917):

> The handicapper (Mr. C. D. Mitchell) evidently estimated the capabilities of the various horses very closely and with admirable results.

Easter Meetings (Narrogin, 1919):

> A Judge's Collapse—When a judge takes a stand it is generally a firm one, but on Monday last Mr. T. A. Hardie, when acting as judge at the Narrogin races, found that although the stand was sufficient to support him and his judgments, when Mr. Starter Mitchell put in an appearance it immediately started to go. Whether Mitch dropped his flag or not was not noticed but the white ants had apparently prepared for a flying start and away went the stand on his arrival. As Charlie also did the handicapping, we think he must have put too much weight on it.
>
> At the Narrogin races on Monday last the starter (Mr. C. D. Mitchell) had occasion to fine a jockey named McDonald for talking whilst the horses were coming up to the starting post. The offending rider, apparently feeling aggrieved at his treatment, made certain unwarranted statements in public afterwards with the result that he was called before the stewards, who after hearing the particulars saw fit to fine McDonald a further £2. There was a very disgruntled jockey in Narrogin that evening.
>
> The close finishes at the Easter meetings which characterised nearly all races was a decided compliment to Mr. Mitchell's efforts to bring the various horses on to an equal footing.

> Charles Mitchell, together with Mr. T. Malony, was handicapper at Wagolin Picnic Race Club's highly successful Annual Meeting conducted on Wednesday 5th March, 1919.

Wickepin Races (New Year's Day, 1919):

> There were several bookmakers, but it is safe to say that none of them went home broke as a result of unduly stretching the odds. Taken all through the handicapping of Mr. C. D. Mitchell was excellent, considering the mixed class of horses with which he had to deal.

Mitchell was vice-president, starter and measurer for the Race Club in 1921, in which year the last two meetings were conducted on the 'Yilliminning Road' course prior to its disposal and the transfer of the Club to the Greater Sports Ground.[7]

The course had the disadvantage of being some miles out of town, with transport not being as freely available as it is today. On one well-advertised occasion, the Race Club hired a train to take the public from Narrogin station to a spot opposite the race track, at one shilling a return ticket; but even this convenience failed to spur the necessary public patronage.

In forthright terms, a report in the *Narrogin Observer* (12 February 1921) deplored the lack of public interest:

> which could incidentally not be ascribed to want of energy or enterprise on the part of the Committee, because this very capable body of men left nothing undone that was likely to make the success of the fixture. If the people will not patronise local race meetings, well the Committee cannot be expected to go out and drag them in by the ears, and that's the end of it.

The committee, taking advantage of the return of horses in the direction of the metropolitan area after a round of racing at various centres south of Narrogin, held a second meeting on Saturday 19 March, 'but owing in all probability to lack of publicity, the fixture failed to attract half as many people as that held six weeks before'.

The Race Club officials were practical minded. To be a success the race course should be in town, and Club officials not only masterminded the Greater Sports Ground scheme, but were its chief supporters. At the annual meeting of the Club on 17 December 1921, Mitchell and J. J. James were elected vice-presidents for the ensuing year. The chairman's report indicated that future race meetings would be conducted on their new course, and continued:

Race meeting, Yilliminning Course

Sir John Forrest and James Mitchell, *c.* 1916

Polo tournament team, 1926

L. to r., Charles Mitchell, Dave Barron, Aubrey Hardie, Jack Brown

Court house, Narrogin
Mitchell's first office

Staff of Agricultural Bank, Narrogin, *c.* 1933
Front row, R. R. Walters, A. E. Illingworth, the author, Shirley East, Thelma Flute, A. N. Treadgold, C. D. Mitchell; *back row,* Evan Davies, William Stratton Ferrier, P. Sanders

Agricultural Bank, Narrogin, *c.* 1933

Available cash in hand has been spent to good advantage, thereby placing the club in possession of a course to be proud of.

The obligation in entering the Greater Sports scheme means that the Club must find at least £75 per annum (rental) more than previously but with the hearty co-operation of past, present and intending members (who will this year be asked to pay a subscription of one guinea) and an increased gate, there should be a prosperous and enjoyable year ahead for both committee and members.

The first meeting on the new course was held on Saturday 4 February 1922, and was both a social and financial success. The *Observer* recorded:

> The course is one of the best appointed we have seen in any country district and the Club is to be congratulated thereon. The stabling accommodation and saddling paddock are all that can be desired. The fields taken all through were very fair but generally speaking the starting was bad, due, in some instances to horses and not to the starter (Mr. C. D. Mitchell) although on one occasion a better start may have been effected. Starting with a flag is unsatisfactory at any time and we understand the Club is installing a starting gate on the course in the near future.

The opposition *Great Southern Leader* remarked: 'Generally speaking the starting was bad, due in many instances to restive horses and not to the starter, Mr. Mitchell.'

With his appointment in 1923 as handicapper for the Great Southern Clubs, his activities as starter and official for country meetings gradually tapered off; but his enthusiasms in Narrogin itself were not curbed until J. J. (Jack) James (president) and Mitchell (official handicapper) were summoned to Perth by the Western Australian Turf Club in the early 1930s to answer charges of sponsoring and taking part in an unauthorized race meeting.

A gymkhana held in aid of a local charity had included in the programme two events for 'Farmers' Hacks' which Mitchell handicapped and started; Jack James acted as judge. Together they had committed the 'grave offence' of not having obtained the Turf Club's official and prior consent. Mitchell told the story of the appearance before the committee.

> We were ushered before the full committee, sitting around a horseshoe-shaped table with Morris Edwards the Chairman[8] looking as bleak as a wet day in the middle of winter. There were no chairs for us to sit down and we remained standing with our hats in our hands. The charge was read and we were asked what we had to say. I'd had enough by this time and told them the meeting was a Picnic one for

> charity and we'd made a good profit, which was more than their damned Club had done, for which we'd worked so hard for nothing. If they had nothing better to do than waste our time, then they'd better come down and take the Club over. With that we stalked out. Poor old Marillier, the Secretary, came running after us, in an effort to sooth our feelings. He asked would we like a couple of tickets for the following day's [Saturday] races. I told him we're staying at the Criterion, send the bloody things around there.

Mitchell very seldom had a bet, and then only 'a couple of bob for an interest'. Betting never appealed to him and 'if indulged in, was likely to upset one's judgement'. He used to boast: 'My only form of gambling is with Nick Anderson. I've had my hair cut there for twenty years and always toss him double or quits.'

He was elected a life member of the Narrogin Race Club at the Annual Meeting held on 12 September 1935, prior to his departure from Narrogin. The gold badge was presented to him two days later when he was farewelled by members at a dinner given for him at the Hordern Hotel.

The chief toast was entrusted to Ted Hardie, who stated that only an event such as this could persuade him to come out of his shell at Albany, into which he had retired nine years before. He continued: 'Regarding the Race Club, so dear to Mitchell's heart, it could be truly said that he had come forward to its assistance when times were bad and the result was shown in the fact that outside of Albany, it was the only live Club along the line.'

W. F. Bird, speaking on behalf of the Greater Sports Ground Board said, 'Mr Mitchell particularly would be hard to replace because his generous spirit and unfailing energy had contributed most markedly to the general advancement of all forms of sport during his long and honourable association with the town. Many of his acts in this regard were unknown to the general public as he was essentially a man who did good by stealth and would blush to have it known.'

Mitchell's only recorded lapse from racing grace occurred in 1927, when he accepted the challenge to lease, train, and race the free-legged square gaiter Trade Wind, a trotter bred by a local man called Bell, out of an unknown mare by Breeze. Members of the family remember having had a great deal of fun out of Mitchell's exercising Trade Wind. Nothing he did could persuade the horse out of the habit of refusing to start but, since Trade Wind had few equals if (and when) he elected to trot, he kept at it. Having brought the horse into racing condition, he entered him in a $1\frac{1}{2}$-mile event in a local gymkhana, but at the last minute he asked W. J. (Billy) Gibson to take his place: because of his own lack of experience in actual racing he considered his driving might prove dangerous to the others.

'Nothing doing, Mr Mitchell', said Billy. 'Dad's driving in this race, and my brother Alan is driving our Roma and she's a goer. You don't expect me to try and beat them, do you?'

Along came their Dad, Jimmy,[9] who said: 'If Mr Mitchell wants you to drive, Billy, you drive and do your best. Give him your instructions, Charlie.'

In those days a trotter's head was held and released on the sound of the gun. Bell, who was as deaf as a post, held Trade Wind, but didn't hear the gun till Billy flicked him with the whip to draw his attention; leaping back, he frightened Trade Wind, who reared and then galloped. By the time Billy pulled him down the field was half a lap in front with Roma 50 yards ahead of the main bunch. Before two laps had gone Trade Wind had caught and passed the main bunch and in spite of Billy's stranglehold on the reins, was hard after Roma.

Billy was still not too keen on beating the other members of his family engaged in the race, and was trying to make Trade Wind break into a gallop and so disqualify himself. His efforts were in vain; Trade Wind had the bit in his teeth and on entering the bell lap went to the front. Chagrin was added to frustration by Trade Wind developing the scours and still winning by twenty yards. Billy recalls that no crowd ever gave a winning driver the equal of the ovation he received that day; it was not for the position he finished in, but the condition.

At the end of Mitchell's lease the Gibsons bought Trade Wind, intrigued by his brilliant though erratic performance and, as they added, 'for their own protection'. They raced him with considerable success for the balance of his career until he was sold as part of the Gibson's milk round, when it was purchased by Mariotts.

Mitchell, on forsaking any financial interest in trotting, at the request of his friend, Jimmy Gibson, accepted the appointment of stipendiary steward to the Trotting Association. Billy remembers him as hard but just. After he had driven one of 'Augy' Trefort's horses at Wickepin in the second race, Mitchell said to him: 'Gibson, I want to see you in the committee room. Don't drive again until I've talked to you.'

'There I waited until after the fifth race; the longest hour and a half in my life', Billy says. 'Then Mitchell dressed me down and gave me a never-to-be-forgotten warning. With Mitchell's judgements there were never any considerations of favours or family influences.'

1. In 1911 the brothers split the Rosedale partnership, Ted taking the western 5,000 acres and naming it 'Denabling'.
2. Guy May a private surveyor, and father of Chas May, later secretary of the Royal Agricultural Society.
3. Sesame was from Grand Flaneur, an unbeaten sire including the winning of the Melbourne Cup in 1880.
4. A. C. N. Olden, of Canadian extraction, erstwhile dentist in Narrogin, volunteered for the 10th Light Horse on the outbreak of war and was accepted as a trooper. He was later promoted lieutenant and attached to 'C' squadron. He received his captaincy and majority whilst serving in the field and on the death of Lt-Col. T. J. Todd succeeded to the command of the regiment. The 10th Light Horse, at 5 a.m. on 1 October 1918, were the first Allied troops to enter Damascus, and Lt-Col. Olden D.S.O. took the surrender of the city from the Emir Said.
5. These short reins are still used by some riders mainly for polo or polocrosse. They are in one piece and of such a length that when held in one hand there is no slack. The grip is enlarged by means of a piece of rubber or leather binding.
6. At time of writing, president of the Narrogin Shire Council.
7. Executive Council approved, on 20 July 1921, the issue of a Crown grant for Reserve No. 7417, in favour of Michael Brown, John Edward Clayton and Alexander Edward Hardie, as trustees for the Narrogin Race Club, on the condition that the land be disposed of and proceeds devoted 'to making a Race Course, Recreation Ground and Sports Ground generally close to Narrogin'. The two blocks, Lots 112 and 164, were sold to Myrtle Mary East of Narrogin on 12 February 1922 for £400.

   Today the land is held by B. and E. Nottle and can be recognized by the line of sugar gums, planted by Mitchell and 'company' on the western boundary of Lot 164 to mark the entrance gates and drive-in to the public viewing space.
8. Morris Edwards, a senior committeeman, was acting in the absence of both the chairman and vice-chairman.
9. The 'father' of the Narrogin Trotting Club, formed in 1921.

14

The Depression

At the peak of Western Australia's land boom there occurred a series of events which reverberated throughout the world. Its chain reaction sounded a death knell to the hopes of many of Western Australia's farmers and, progressively, to those of the country storekeeper and to an adverse degree of the whole of Australian society.

F. K. Crowley, in his *Australia's Western Third,*[1] summed up that position so concisely that it is worth enquiring if forty years later history might not repeat itself. 'The year 1929 saw the end of the experimental period of agricultural development and showed that farmers were by no means masters of their own destiny, hard work was not all that was required for success.'

Thursday 24 October 1929 is the first of the days which history identifies with the onset of the Great Depression. The venue was New York; the scene the Stock Exchange; the occasion, *Black Thursday*; the catalyst, panic selling.

There had been Stock Exchange panics and violent market fluctuations before. A common feature of most earlier troubles was that, having happened, they were soon over. The worst was reasonably recognizable as such. The singular, and for a time unrecognizable, feature of the 1929 crash was that the worst continued to worsen.

The *West Australian* of Friday 25 October gave the panic a passing news mention in the trade and finance section (p. 19) :

> New York, Wednesday: Chicago Wheat Prices slump sensationally today with a bear stampede, following the receipt of news of declines on the New York Stock Market and the lack of an export demand.

On the following day, again under the trade and finance section, the *West Australian* reported:

> Winnipeg, Thursday: Wheat tumbled 13 cents a bushel during early trading today. Bearish news from Europe and the Argentines and weakness on Wall Street being contributory causes.

During the following week the price of Western Australian wheat declined two shillings a quarter.[1a] The worsening share market scare spread to London, but Australians believed there was no need to concern themselves unduly. New York's financial panic was brushed aside, the basic causes being described, at Australian level, as purely technical and psychological—a method of getting rid of the lunatic fringe of American investors, those who traded in margins.

Wheat continued to drop but 'what the hell?' Didn't Australia produce the world's best wool and wheat, and didn't the people of the world need to feed and clothe themselves? Australians had more to concern them, gripped deeply as they were in the throes of their annual fever of the Melbourne Cup; all else must wait until Tuesday 5 November, when the favourite Phar Lap would be beaten by Nightmarch.

There had been earlier warnings. At the wool sales held in the previous July the principal brokers, Elder Smith and Co., Dalgety's, and Goldsbrough Mort and Co., had offered a total of 280,000 bales. Only 198,000 were sold, at prices ranging from 5 to 15 per cent below the previous average. Wool for the 1929-30 season f.o.b. averaged 10.53 pence per lb against 16.26 pence for 1928-29.

It was in July also that London's financial houses refused any further long-term loans to Australian governments.

The New York Exchange's trading continued to decline until it reached the all-time low of November 11. The closing of the doors of prominent business institutions in the United States and details of suicides among prominent financiers made the main headlines.

At the Annual Meeting of shareholders of the English, Scottish and Australian Bank in London, on November 13, the chairman of directors said that 'after many years of great prosperity, stimulated by a large inflow of borrowed money, Australia was passing through some difficult times'.

But this item was reported only in the *West Australian*'s trade and finance section, and who bothered to read that? A. H. Braine, secretary of the Co-operative Wheat Pool of Western Australia, announced in the November 1 issue of the *West Australian* that the First Advance on pooled wheat would be 3s 6d per bushel. This advance would represent 78 per cent of current market value and was 2d per bushel more than the First Advance at the same time in the previous year.

At the opening of the 1929-30 season in October wheat was quoted at 4s 5½d at siding, on 4d freight basis. By the end of February the price had dropped to 4s 0½d. Many growers who had taken First Advances of up to 4s from the private buyers continued to hold out for a rise in price, until they were sold out in December 1930. The amounts they had been over-advanced took years to repay.

In February 1930 the Prime Minister of Australia, the Honourable J. H. Scullin, urged wheat growers to 'Grow More Wheat', and so maintain the total value of wheat exports—a very necessary admonition since Australia's balance of payments was in a state of acute strain. Those who still remained on their farms, and who could obtain the necessary credit, responded to the appeal, and Western Australia's Director of Agriculture, G. L. Sutton, stumped the agricultural areas in active support. The area of the Australian wheat crop sown was 3,000,000 acres greater than in the 1929-30 season, and production was 87 million bushels above the previous year. In Western Australia the increases were 388,000 acres and 14.4 million bushels.[2]

Undoubtedly at the time the Prime Minister was sincere in contemplating a guaranteed price of 4s a bushel. Legislation to that effect was introduced later in the year, but as by that time it had become patent that a price slide was well under way, the bill failed to pass the Senate.

A. J. Monger, chairman of the trustees of the Co-operative Wheat Pool of Western Australia, stated in the *West Australian* (16 October 1930) that it was expected that deliveries of new season's wheat would commence that day. The trustees of the Wheat Pool, assuming that the value of 28s 7½d per quarter, and the average freight shipping rate of 35s per ton would be maintained, felt that the value of wheat could be 2s 9¼d f.o.b. The Pool had arranged with its bankers not to draw more than 80 per cent of the f.o.b. value and it was seen that on an average siding rate of 4⅝d the First Advance should equal 1s 7d.

However, advice of the First Advance was held over until November 1. On that day, the First Advance was fixed at 1s 4d. A. J. Monger stated that 'he feared that the public would realize for the first time how really serious was the position for Western Australia'.[3]

This announcement, however, did not evoke the scare headlines one would have expected. Were we, as a state, just numbed, or fatalistic? Unemployment had doubled, wool continued its downward spiral, new bank loans had become the exception. Once again Australians were more concerned about the Melbourne Cup, which was to be run three days later, and which was to be won by the public favourite, Phar Lap, from Second Wind. The federal government withheld its budget until after the race had been run, but on the following day announced the first of the 'gloomy, belt-tightening' budgets.

The Acting Treasurer, J. Lyons, budgeted for a deficit of £8m to be found mainly from increased taxation (£3.7m) and various economies estimated at £3.18m. Although unemployment had continued to worsen—the 1929 figures had doubled by the end of 1930[4]—top level recognition of the position had to wait until now, the publication of the budget, for the first scare headlines.

The governing Labor Party split in two as Caucus recommended that Australia should default in respect of payment of a loan of £27,000,000 then falling due, suggesting that payment be deferred for twelve months. Worse was to come. In February 1931 Australia's currency was devalued to £A130 = £E100, although in December of the same year it was slightly revalued to £A125 = £E100.

By mid-1933 Australia had hit rock bottom, with the 1929 basic wage of £4 17 0d reduced to the all-time low of £3 13s 6d, and one-third of Australia's bread-winners out of work.

In the years following 1930-31, although wheat prices remained very low, production continued at an astonishingly high level. With reduced average returns per bushel, farmers extended production in an attempt to either maintain money incomes or cover fixed interest charges (which naturally came more and more into line with net incomes as the latter fell).

Interest rates for Agricultural Bank loans fell as follows: From 7 to 6 per cent as from 1 January 1931; to 5½ per cent as from 1 July 1932; to 5 per cent as from 1 January 1933; Discharged Soldiers Account: From 6½ to 5½ per cent from 1 January 1931; 5½ to 4½ per cent as from 1 August 1931; and were followed in similar ratio by private banks and lending institutions.

By mid-1930 people began leaving their farms. There had been little warning of disaster—neither the state nor the farmers had capital reserves. Hope died quickly, as there was no-one in authority who could offer it with conviction. Despair descended on the lately settled agricultural communities, where the general reaction was to leave the farms and to return to the cities, in the hope of finding employment. Men became the responsibility of the government there; a fixed wage or sustenance grant became their idea of security. The majority of those who remained lived to benefit from the decisions even though it meant waiting until after World War II, but during the early years of the Depression they went through extreme hardship.

I know of no recorded deaths from starvation in the wheatbelt areas during the depression years, but I am certain that the philosopher who coined the phrase: 'Hunger is a good sauce'—never experienced the monotonous and unattractive fare which for four years was usually the farmer's lot. The staple dish was boiled wheat and treacle—known as 'cocky's joy'—with rabbit, kangaroo, or emu to supplement the ration. Golden syrup, 3d more for a 2 lb tin, replaced treacle as a special treat for Christmas.

Never-the-less, many pleasing dishes were devised by the ingenious farm house-wives—lightly browned wheat porridge, for one. Several handfuls of wheat tossed by hand in a dry pan over the fire each evening, then put through a mincer and placed in a saucepan and covered with water, produced on the following morning a tasty porridge which required no sweetening. Sugar, of course, was a luxury item, and if there was no house cow a tin of condensed milk went a long way.

J. B. (Jack) Curlewis, who farmed at Dowerin from 1927 to 1931 before joining the Agricultural Bank as an inspector, recalls how he and his wife lived on a pound's worth of groceries a week, which comprised mostly baby food, flour and salt. Their staple diet was 'cocky's joy', supplemented with rabbits and fowls. As a relief from these dishes, there were also eggs, fresh or preserved—it was easy to feed fowls in those days. Mrs Curlewis baked wheat in the oven and put it through a mincer to produce an excellent substitute for coffee.

One of the great delicacies was a pie of galahs—the grey and pink variety. The birds were trapped with a trail of wheat leading under a box supported by a stick which was attached to a long string and yanked away at the appropriate time. Fortuitously, also, 1930 was a phenomenal year for mushrooms, and although enough ketchup was bottled to last for eighteen months, the delicious sauce has probably never been served with such uninteresting food.

The Curlewis conveyance was a Model T Ford, which was used also to pull a trailer with water for the house. When the clutch lining gave out, it was replaced with Jack's leather belt, but he was then no longer game to chance bogging and burning out the clutch plate forever. During the last twelve months on the property all house water was carried from a soak 15 chains away in two four-gallon kerosene tins on a yoke. There was no money to replace inner tubes when they wore out or blew out; in their place the tyre was packed tight with dried horse manure and by the time this substitute was common to all four wheels, the ride, reportedly, was quite smooth. Jack is of the opinion that this substitute was possible only because of the spring steel tyre rims which were the fashion of those days.

Socializing went out of vogue as the last half of 1930 came in. While you were still welcome to make calls on your friends and to share what little there was, you brought your own beer, if you could afford it. Drinking in hotels and clubs continued as a male privilege, to be preserved as long as spiritous beverages are brewed. But 'shouting' was taboo; 'a shilling in' became the rule. As one never drank unless one was thirsty, pots were the order of the day. These held 14 oz in the early days of the Depression, but down to 12 oz by the end. Drinking a glass was regarded as a waste of money. 'Sixpence in only' was also acceptable, if it either fitted the person's pocket or capacity.

At the time of the Depression there were displayed at vantage points throughout the wheatbelt two distinctive sheet-metal signs, each having a blue enamel surface, advertising in white letters a very prominent insurance company or a well-known brand of windmills. It was not long before someone discovered that these signs not only made excellent stove tops, but also, when built into mud-bat fireplaces, oven sheets for baking bread. I cannot remember having seen one in its rightful display position after my return from the war, and can now confess (seeing that the Statute of Limitations has expired) that I carried one for three years myself! It proved excellent as a top for the camp fire during the winter, and a windbreak for the fire during the summer.

The economies forced on farmers, other than that of sustenance, were severe. Many were compelled to resort once again to the use of horses, while to reduce the fuel costs of kerosene-run tractors, others switched to the use of producer gas made from charcoal. An excellent charcoal was produced from mallee roots, cleaned of soil through exposure to the elements, and burnt in a large pit, covered with sheets of roofing iron.

Economies could not always be made, however, and in many cases inexperience, bitterness or lack of spirit precluded their being made. In many instances settlers just walked off their farms, some even without notice to their inspector or branch office, although in such cases there was no equity for those who had given their all, and little though that may have amounted to in terms of money, almost invariably it represented the loss of the best years of their lives. The bank staffs worked all hours: stock had to be cared for, particularly horse teams; machinery needed to be stored, buildings to be preserved.

For the years ending 30 June 1930 and 1931 the trustees of the Agricultural Bank recorded in their Annual Reports:

> Reverted properties on hand, either from default or walking off total 493, 336 being in the Wheat Belt and Great Southern areas and 157 in the South-West and metropolitan/suburban. This high number was in spite of 252 properties sold during the year, with a bank loss of £52,445 which was written off.
>
> At the end of 30th June 1931, 631 reverted properties were under the control of the Bank, an increase of 138 on the previous year. This total was exclusive of a further 226 properties sold, principally to adjoining holders, with a recorded loss to the bank of £38,896.

MITCHELL TOOK THE FARMERS' PLIGHT deeply to heart, considering it much more devastating than the $22\frac{1}{2}$ per cent cut he had suffered in his own salary.[5] Fortunately he was able to treat the situation objectively; while feeling for all, he did not let himself become emotionally involved over the fate of any individual. The farmers were still his responsibility as a body, to be protected from themselves and from creditors, although he was ever mindful of friendship and long association, sharing with all the pervading deep despair; he never forgot that his duty lay with the majority.

With state expenditure curtailed to a bare minimum,[6] both in work force and travelling expenses, Mitchell and his fellow field inspectors became the 'work horses' of the government departments—handling all Lands Department inspections and reports, paying not only the unemployment relief to families placed on vacant properties as caretakers, but also the subsidies provided to those prospecting for gold. These and a thousand other responsibilities contributed cruelly to Mitchell's administrative load and worries, but to him and his team no task was too menial and no decision was shirked.

At the height of the Depression and a time of universal salary cuts, to offset their extra duties inspectors' fixed travelling allowances were increased from £150 to £175 per year for those with cars and from £125 to £150 for those with horse transport. Regarding the latter, Inspector A. A. Mackie of Karlgarin is the only one I can recall who still used a horse and sulky; when he left on a trip of inspection he was lost to all officialdom for about three weeks, to its great annoyance and confusion, and, no doubt, to his own peace of mind.

Two problems harassed the Bank exceedingly. Credit available to farmers for seasonal requirements was now the exception rather than the rule of the previous ten years, and the Bank, to provide or arrange even these small disbursements, had to survive. It could continue to function only if it could be assured of its 'life's blood'—payment of current interest. From this latter source the Bank had to provide for the servicing of its borrowings, for staff salaries and for administration costs. Its problem was, first, to provide farmers with superphosphate and sustenance for seasonal cropping and, second, somehow to obtain payment of at least current interest on loans.

Those settlers who were still under the aegis of the Industries Assistance Board and who had made sufficient repayment from the previous season, could rely on assistance being forthcoming for the next crop. So, to provide a similar measure of security for the balance of the Bank's clients, Section 37 (a) of the Agricultural Bank Act (No. 15 of 1930) was passed and assented to on 19 November 1930. This section provided for money due for fertilizers and interest to be charged on land, crops, etc., and read:

> (1) In respect of—
> (a) any advance made at any time hereafter by the Bank to any person for fallowing, fertilisers, insurance of crops or bags; or
> (b) any interest for not more than one year which has become payable to the Bank by any person on any account,
> the Bank shall be entitled to a first charge, in priority to all other encumbrances, upon all crops sown or to be sown upon any lands of any tenure held or occupied and used by such person for agricultural or farming purposes.

Section 37 (a) went on to provide for the Bank on written request of merchants supplying fertilizer or bags on credit being granted priority for their supply to the Bank's charge for interest.

But in granting this priority the Bank did incur any liability should the crop fail or proceeds be diverted.

With the proceeds of the majority of crops being now automatically secured to some degree or other, the farmers looked for some avenue to divert a proportion of the proceeds to their own private use. Then, misappropriations (or stealing) of some part of their crops were neither legal nor moral, but to Australians, with their delightful flair for fitting the word to the action, the practice became known as 'dummying' or 'flogging'—a 'lurk' to play at so long as you were not caught! Those were difficult times and one must not judge too harshly. It was a 'game' of personal necessity played, to some degree at least, by most farmers. It was a protest against entrenched bureaucracy, and its lack of understanding, against a breakdown in communication. It was no felony unless one was caught, and even a conviction carried no social stigma.

Creditors were far removed from the actual scene of operations, and their appraisement was hard and cold. Mitchell, on the spot and hot, expressed a point of view that might not have been endorsed by officialdom, but which was at least founded in sympathy and understanding. To enable him to dispose of whatever part of his harvest the farmer might dummy on his own behalf, the local storekeeper would often acquire a nearby property and sow a crop on it, thus to a degree legalizing and certainly justifying the sale of wheat in his own name. Some of them, however, rather over-gilded the lily. I well remember Mitchell standing in the main street of a small country town, and telling off a storekeeper who had complained that he was being harassed by Mitchell's staff under direction.

'Look here', Mitchell roared. 'I don't mind your being a rogue and shutting my eyes to your marketing a 60-bushel crop. You're entitled to some return for helping the farmers over the past twelve months. But only a thief and a bloody liar would expect me to accept a claim for a 60-*bag* crop!' A 60-bag crop meant 180 bushels to the acre, when the average yield that year, in that district, was 12½ bushels.

There were other means, normally regarded as much safer, but I recall also one such dodge which didn't turn out to be as safe as had been thought.

There actually was in those days a farmer whose property was not under mortgage, so he was at liberty to dispose of his crop at will. He was approached by practically every farmer in the district with a request to sell a number of bushels in his name. Kind heartedly, he agreed. Between delivery of the various parcels of wheat and receipt of payment, the kindly farmer had a sudden heart attack and died. His executor, and principal beneficiary, a nephew from the city, found it hard to accept the unrecorded claims of those who had sold wheat in his uncle's name and now expected payment. There was no evidence beyond word of mouth and, peculiarly, the claims of the neighbours exceeded the total amount of wheat marketed in the uncle's name, including his own crop! The incident was rather Gilbertian: the claimants could neither sue nor give any publicity to the refusal of payment. Mitchell and his team chuckled; this was mete justice!

One might gather the idea that the principal buyers of wheat were party to the 'flogging' trade. Not so, but they were not so simple as not to know what was going on. Storekeepers, who may have had an axe to grind, and wheat lumpers, in charge of deliveries, who knew and sympathized with the farmer, were the 'on-the-spot' agents.

Necessity has always been the mother of invention and in these lean years, practically overnight, pig farming blossomed into a profitable sideline. The 'missus' pin money' airily dispelled the importance of the operation. To some farmers it became their life-line. They could avoid delivering wheat to their creditors by walking it off their properties as pork or bacon.

This sideline was winked at; it kept the farmers happy, but I doubt if anyone in authority ever realized until after the war years the tremendous total profits that resulted from pig raising. Baconers were sold direct to processing firms, over scales, at so much a pound. Cheques were written on the spot by the buyer and Nelson himself could not have turned a more blind eye on the names given by the vendors. As a country branch manager after World War II, I was amazed for a while at the number of 'pig cheques' cashed in the name of 'Bill Smith'. The 'Bill Smiths' numbered slightly more than the 'Bob Menzies'. There was no lack of pseudonyms to choose from, appropriate to the occasion. Farmers had their own sense of humour, neither vicious nor intentionally hurtful, but certainly subtle. Ron Richards who, some years later, followed me along as manager of the same branch, was amazed one day to open a letter addressed to him from the Department of Agriculture, which said:

> The pigs you sold at last Thursday's market were found, on arrival at Fremantle, to be heavily infected with lice. Please ensure that the following control measures are carried out before any further pigs are marketed.

It was the joke of the district—the actual joker still unconfessed but suspect.

To curb the practice of flogging and certainly to keep it within reasonable bounds, the Agricultural Bank employed as its detective Thomas Stephen Hickey, who had retired in 1909 from the C.I.B. with the rank of detective, to go farming at Trayning. Tom, of medium build, thick set, with a revolver swinging at his hip and a cigar perpetually rolling between his teeth, looked the part of a Mid-West sheriff. He was rough and tough. A farmer seated on his tractor being quizzed by Tom lost his temper: 'Nobody's going to talk to me like that, Hickey, I've a good mind to get down and give you a punch on the nose!'

'O.K.', replied the unperturbed Tom. 'Hop to it—that's what I'm paid for.'

However, there were more frights and warnings than convictions: it was an unpopular but effective appointment; creditors generally were satisfied that their interests were being protected.

Wheat was bartered with anyone who would accept delivery in exchange for goods. Apart from storekeepers, garages, hotels and even tobacconist-hairdressers accepted the system at prices from 5s to 6s a bag. Many a time during the hot and dry wheat-carting season I saw a carter put a bag on his shoulders and dump it on the pub counter, with a 'Cut this out'. It was then the responsibility of the recipient, in his turn, to find some means of flogging wheat, but it is not the responsibility of this story to recount the means employed!

The depression years witnessed days of quick decisions and of grasping at every straw. The Finance and Development Board Act No. 47 of 1930 ('an Act to provide for the establishment of a Finance and Development Board for Western Australia to define the powers and functions thereof and for other incidental purposes') was assented to on 22 December 1930. The second reading, as reported by W.A. Hansard, records:

> This Bill is intended to, and will, if passed, give authority to a Finance Board to provide funds for the Agricultural Bank. Parliament has always appropriated funds for this institution and today we are depending upon the same method for the raising of money by the Federal Treasurer.

> The Financial Agreement now controlled the raising of loans; but in the other States there were semi-Government institutions which have separate borrowing powers and this Bill sought similar powers for the Agricultural Bank. The Bank must be regarded as somewhat apart from ordinary Government activities and carrying responsibilities which must be met.[7]

The Finance and Development Board was established, comprising the managing trustee of the Bank (who was to be chairman) the Under Treasurer and the Public Service Commissioner. The first two were appointed ex officio, the third by the government.

After the act came into force no moneys were to be raised by mortgage bonds as provided in the Agricultural Bank Act of 1906. The Board, with the approval of the Governor, was empowered to borrow money for the purpose of making advances to the Bank, and Section 14 provided that 'all moneys borrowed by the Board may be raised by the issue of debentures, payable to bearer with or without interest coupons attached, and charged and secured upon the income, assets and other property of the Board'. All funds then secured to the trustees of the Bank were vested in and transferred to the control of the Board.

The Farmers Disabilities Royal Commission report in the following year, was most critical of the Board's composition:

> It is difficult to conceive at the present juncture of affairs in Western Australia what time the Under Treasurer and the Public Service Commissioner of the State can afford to devote to the affairs of the Agricultural Bank, Soldiers Settlement, Group Settlement and Industries Assistance Board.

The harsh criticism of the Royal Commissioners appears unjustified. In their first year of operation and prior to the issue of the Royal Commission's report, the Finance Board borrowed £500,000 on Debentures, which apparently proved sufficient to meet the balance of the farming clients' cropping and carry-on needs.

Following the failure of the federal government's proposed bill for the guaranteed price of 4s for the 1930-31 wheat crop, it had become increasingly apparent that both secured and unsecured creditors were going to wait no longer than the current crop to collect what they could of their dues—literally 'hop in for their chop' and Heaven help the debtor farmer. To prevent the anticipated chaotic rush of bankers, traders, merchants, agents and businessmen hastily trying to collect their debts, the government obtained approval to introduce a moratorium intended to protect all farmers from immediate foreclosure or from being sold up.

The Farmers' Debts Adjustment Act (No. 49 of 1930), assented to on 31 December 1930, was a matter of haste and good intention, but based more on expedience than experience. It was amended quickly and considerably by Act No. 4 of 1931, assented to on 10 July 1931.

The act had a mixed reception in the Legislative Council; some members were highly critical. The Select Committee of the Legislative Council dealt with the bill in its first form: it was stated that the bill already had a damaging effect on credit, and that, if it were placed on the Statute Book, there was reason to believe 'it would dry up the stream of credit in the State, because no person would feel inclined to advance money when he was likely to have his debts frozen'.

Under the terms of the Farmers' Debts Adjustment Act any farmer, or the creditor of any farmer, could apply to the director to call a meeting of the farmer's creditors. The director (W. A. White) could thereon grant the farmer a stay order as a protection against creditors, pending examination of his affairs. After claims and statement of assets had been examined (in the first instance by officers appointed either from government sources or Agricultural Bank staff), it would then be decided at a subsequent meeting of creditors how many or how much of the debts were to be written off and how much each farmer would have to pay while still working.

A proposal for carrying on and financing the current season's operations was placed before creditors. If the creditors were agreeable that their claim should remain in abeyance for another season and were prepared to assist the farmer by means of cash or goods, the claim for carry-on was adopted, and a receiver was appointed to control the farmer's expenditure within the limit of the creditor's decision.

At the end of the season all the creditors were called together to adopt the receiver's statement and to observe the result of the year's working. If the farmer had given satisfactory results for the season, the same procedure was followed in order to carry him on for a further year.

There was no fund in support of this act, and its viability relied purely on the assistance of creditors. A simple majority in value and number of those creditors present validated a resolution in favour of or against the farmer.

The act was, perforce, a purely lubricative affair. It certainly helped those who merited help, but omitted those who needed help the most.[8]

Overnight Mitchell found himself appointed receiver for thirty accounts. He never for one moment shirked or endeavoured to get rid of the responsibility. Whilst his staff maintained the records, Mitchell supervised and authorized every transaction—no longer were there spare weekends. Here it was that I glimpsed at first hand Mitchell's complete grasp of farming essentials—the minimum required for economic and efficient husbandry; the minimum required, not necessarily to recoup the farmer for his labour but, at least, to maintain his dignity. What was left over was meticulously divided. He seemed to know instinctively, and to treat as though they had never been employed, the subterfuges employed by both farmer and creditor to get a greater share.

By the end of 1931, when by official instruction all F.D.A. accounts were turned over to private enterprise, I had absorbed the first principles of farm management—a subject that has stayed and progressed with me throughout my official life.

As I have stated, the moratorium created by the Farmers' Debts Adjustment Act gave protection to only a small proportion of the farmers in need of it. Evictions and re-possessions of plant and machinery were still the lot of the unfortunate majority. Naturally, there was resistance at both personal and organized district level. Those persons acting as bailiffs almost always sought police protection, and the police possessed a keen sense of justice.

The bailiff or representative presenting the notice for repossession of a harvester or plough would at times find two or three neighbouring farmers also waiting for him when he arrived at the property concerned. The debtor would yield possession, but before the machine could be moved one of his friends would claim some part of the machine, usually the main wheel, which he stated he had lent to the farmer. By the time each of the neighbours had demanded return of some part which each declared had been lent, it had become practically impossible to move the machine economically. It then became a question of whether leaving it on the property and chancing eventual payment might not be a more attractive proposition, practically, than transporting and refitting it.

Stories of the attempted re-possession of tractors are legion, but my favourite concerns a bailiff who arrived with the local constable to find the tractor in the middle of a paddock and surrounded by some twenty farmers, each carrying a rifle or gun under his arm. Between them and the tractor was a circle of bottles, and while the visitors walked towards the group, occasional shots rang out and bottles shattered. The policeman queried this apparent armed resistance, but was informed that it was merely a practice shoot of the local Gun Club which had coincidentally leased from the farmer the particular piece of ground on which the tractor stood.

'Well', said the policeman, turning to the bailiff, 'of course if the Gun Club has leased this land for target practice, I don't know where you stand in regard to the hazard of crossing their line of fire—perhaps you had better wait until they have finished practice.'

On enquiry it was found that the 'Gun Club' was on a two-day shoot which had only just commenced. As the bailiff hesitated, a bottle was shattered just in front of him; he jumped and another one slightly to his right went off. He decided that discretion was the better part of valour, and left the scene of battle.

One of the group, telling the story afterwards, said: 'We were only bluffing, but if he had walked through the line of bottles, we had a man ready to jump up and sit behind the wheel of the tractor. He would have refused to move, and if the bailiff had tried to shift him by force, we would have hit him with a summons for assault. We had it covered from all angles.'

James Bligh, a volatile Irishman of whom we read again later, had some bother early in 1937. His harvest had averaged only 4 bushels per acre and he was unable to meet payments on the harvester, drill and truck. McKay Massey Harris Pty Ltd, which owned the harvester and drill, was content to stand payment over till the next year. Although Bligh had managed to scrape up £90 towards part payment of the promissory note on the truck, the company stated its intention to re-possess it and then re-sell it to Bligh under a new agreement. When the agent came on to the property, Jim was very upset to learn that the company had added £41 worth of 'various charges' to the price of the truck. He refused to sign the document, or to allow them to take possession of the truck. To his plea: 'You're robbing me!', the representative, according to Jim, said: 'Well, we'll see about that', and motioned with his hand towards his hip pocket. Jim said he didn't know whether it was to pull a revolver or not, but without waiting to see, took up his rifle and placed a shot into the ground about

3 feet in front of the agent. With that, the chap took off. For good measure Jim put another shot behind him and another one to his right. Jack Robb, the farm hand, mistaking Jim's wide shooting for poor marksmanship, grew excited and yelled: 'Here, boss, give me the rifle, I won't miss the b——.' Where, today, would you buy loyalty such as that for £6 a month and board?

DURING THE DIFFICULT TIMES OF 1930 AND 1931 there was no-one who could speak with authority and from experience as to what to do next. Advice was not in short demand and proffered freely. Opinions were as varied as the advisers and unsolicited critics.

From this maelstrom of 'thrashing about' for answers there emerged, as a seeming matter of course, a pattern of the times, the inevitable face-saving or buck-passing Commission of Enquiry into the matter of liquidating farmers' debts and the disabilities affecting agricultural industries. The Farmers Disabilities Royal Commission was commissioned by His Excellency John Alfred Northmore, administrator, on 26 May 1931. Members consisted of A. H. Dickson (chairman), Harry Hale, G. H. Carlisle, Theodore Pellow and F. C. Farrall. R. W. Brownlie, seconded from the Treasury Department, was secretary.

The Commissioners were empowered:

1. To enquire into, investigate, and make recommendations upon any proposals brought before the Commission for the liquidation of farmers' debts, and in particular upon certain schemes for such purpose already submitted to the Government of the State.
2. To enquire into the disabilities affecting the agricultural industry in the State, and in particular to report on the main causes affecting the cost of production, including interest charges, the burden of the Customs tariff, and all other factors.
3. To make recommendations generally in respect of the matters investigated, including the recommendation of such legislation, as are deemed necessary or expedient to carry into effect such recommendation as we, your Commissioners, may make under the authority of the Commission.

The Commissioners reported subsequently that they had restricted their enquiries to the following classes of farmers: (*a*) wheat alone, (*b*) wheat and sheep, and (*c*) grazing, comprising pasture, oats and sheep.

Considering the importance and depth of the enquiry, this Commission must have established at least a Western Australian record in as much as that the report, dated 31 July 1931, was completed and submitted within sixty-six days.[9] There was need for haste, but one cannot help but wonder now, in retrospect, if there might not have been some degree of prior direction or the undue application of personal convictions.

The conclusion of the Commissioners read:

> Your Commissioners desire to impress upon all sections of the community their dependence directly or indirectly on primary production. Anything which hampers primary production has its corresponding effect on the general interests. If the primary producer fails, it is reflected in the state of trade generally, the incapacity to pay wages and salaries, in city land values, and in many other ways. It therefore behoves all sections of the community to realise that for the preservation of the State and in the interests of all, the farmer must be given a reasonable chance to obtain some compensation for his labours, and when through the collapse of markets over which he has no control, the farmer should be free from the fear of losing the heritage he has toiled to create for his children. Your Commissioners are convinced that reductions in costs are more permanently reliable than dependence on the vagaries of world's markets. Further, your Commissioners would impress upon the members of our Legislative that 'time is the very essence of the contract' and that financial preparations must be instituted forthwith for the harvesting of 1931-32 and 1932-33, so that there will be no heart-breaking, harrassing, and unnecessary delays in bringing forward the required legislation.[10]

The report appeared to be weighted in favour of maintaining the status quo among wheat farmers by providing finance for taking off the 1931-32 harvest and putting in and taking off the 1932-33 crops—purposes which reflected support for those of the merchants of the state who, in particular, benefited by and from primary industries.

This impression is strengthened by the minority report submitted by F. C. Farrall, and paragraphs 2 and 3 of his report are quoted:

> The plan submitted makes full and ample provision for carrying on settlers who are in difficulties, but there is no recommendation suggesting any relief whatsoever to the wheatgrower who, though able to arrange his own finance, is nevertheless producing at a very heavy loss.
>
> The evidence submitted to your Commission proves that wheatgrowing cannot be continued on present prices. This fact is fully borne out in the majority report, and unless some provision is made to ensure that wheatgrowers are recouped at least the cost of production, a state of collapse must assuredly follow.

Mitchell was among the six Agricultural Bank officials, including the managing trustee, A. E. McLarty, who gave evidence. No record of their depositions can be traced, but one statement by McLarty was recorded, and is well worth repeating: *'There is also much bad farming going on. We have some of the worst farmers in the world.'* The report generally was highly critical of Agricultural Bank administration, stressing that the managing trustee required far more assistance than the part-time services of the two co-trustees. The remuneration paid to these two trustees was five guineas each for sitting and the maximum to be paid to each trustee was fixed at £250 a year.

The report went on further to state that the two co-trustees had outside interests to which the major part of their time was devoted. Criticism of the managing trustee ('who has been overloaded with work. Not only is he Managing Trustee of the Bank, but also of the Soldiers Settlement, Group Settlement, and the Industries Assistance Board') was realistic and constructive.

Several of the Commission's other findings brought sharp retorts. An extract from the Annual Report of the Agricultural Bank for the year ended 30 June 1931 read:

> The Trustees have read the report of the Commission appointed to enquire into farmers' disabilities and have noted its comment regarding the activities and administration of the Bank, particularly in regard to its failure to collect overdue debts.
>
> The Trustees have not seen the evidence on which the Commission based its findings, but it has to be considered that the Bank in the past has had no means of controlling the crops and other incomes of its mortgagors, having in case of default to rely entirely upon the exercise of its power of sale; but with a constant reversion in considerable numbers of weakly held securities this is a remedy which even under normal conditions must necessarily be used with moderation. An amendment of its statutes in November 1930 (Section 37A) entitles the Bank to a prior charge over the crops of its mortgagors to the extent of one year's overdue interest. This, however, to a very great extent, has been defeated by the refusal of the trading community to provide essential credit except on an undertaking by the Bank to postpone its charge. Many wheat growers are experiencing trouble in procuring assistance to harvest and market their crops, with the possibility of credit being no easier to obtain for the ensuing year.

S. B. Donovan, in his minority report of the Agricultural Bank 1933-34, remarked:

> The Farmers' Disabilities Commission in 1931, after taking evidence from farmers other than Agricultural Bank clients, realised the hope-

lessness of the farmers' position, and held out the 'olive branch' to the mortgagees, creditors, secured and unsecured, to come together, accept the inevitable, and voluntarily agree to an equitable adjustment of farmers' debts.

Donovan commented further: 'A policy that merely aims at keeping the farmer on the land cannot be pursued indefinitely.'

The report would appear to have had neither legislative nor administrative effect, but perhaps these aspects may deserve a closer scrutiny than has been given here. Possibly its influence lay in pointing the need for re-organization within the Agricultural Bank, which was given effect to by the Labor government succeeding that of Sir James Mitchell on 24 April 1933.

However, the winds of change were blowing, not strongly, but in the direction of relief to Western Australia generally and to an improvement in status and autonomy for the Agricultural Bank's district officers.

The year 1933 saw the establishment of the Commonwealth Grants Commission, aimed at preventing the poorer states—Western Australia, South Australia and Tasmania—from becoming bankrupt.[11] Thereafter these states (later known as the 'mendicant states') received realistic annual grants from the Commonwealth.

The price of wool rose sharply at the opening sales and demand held for the 1933-34 season, yielding an average price f.o.b. of 15.65 pence per lb, nearly double that of the previous year. Hope rose proportionately.

Two actions gave the Bank new vigour. Group Settlement had been transferred to the control of the Agricultural Bank as from 1 October 1930, which action brought in its train the transfer of R. G. Courtenay (at the request of the government) from the Lands and Surveys Department as chief accountant on 1 August 1931. An able and dispassionate administrator, Courtenay became the strongest single force in the Bank for the next three and a half years. Things began to move at last, and with the transfer of W. H. Cover from Narrogin to Head Office on 26 May 1931, the re-organization began to have practical results.

A selfless, experienced and sympathetic officer, who held the confidence of branch staffs, Cover was initially employed as the Bank's representative at creditors' meetings called under the Farmers' Debts Adjustment Act. He became Courtenay's right-hand man with his appointment to the position of assistant accountant and inspector, as from 18 May 1932.

After its initial labour pains in the early 1920s decentralization had long since become static in the non-relaxation of Head Office control and discouragement of on-the-spot decision making. Now it again became a positive practice.

As from 1 July 1932 the archaic titles of senior clerk and paying and receiving officer were amended to accountant and senior clerk, divorcing the positions from the time-worn Civil Service nomenclature and relating them to those of similar responsibility in the trading banks. The moves, insignificant as they may appear, were worthy of editorial comment in the *Sunday Times,* 3 July 1932.[12]

I have described this chapter as depressing, and in many ways it is. One cannot recall, and write about, those stringent times without experiencing once again, even at this distant remove, something of the agony of despair which gripped those unfortunates who had no regular income. But on further consideration, and now that one's memory is stirred by recollection also of some of the men and women one encountered, and of the remarkable courage and spirit and resilience they somehow dredged up to face the interminable work and worry of those lean years, it is with something approaching wonder that one remembers how much humour, and excitement and interest was woven into the sombre woof. I don't suggest for one moment that the qualities which brought the people of this state through that testing time are peculiarly ours, but I do own up to considerable pride in the fact that they were evident in such abundance.

To me, Charlie Mitchell seemed, then and now, to be the quintessence of those men—mostly the unsung heroes—who braced their backs and pulled toward recovery. He worked then as I have never seen any man work before or since, and by his own example lit a fire under scores of others who might well have become indifferent to the course of events.

I think again of Mitchell—the gruff, the shy, the soft-hearted—the man of unplumbed depths. In doing so I recall the day, in the late summer of 1933, when we called at a farm 29 miles from the nearest siding—10 miles from the nearest neighbour. The couple, with three young children, without any form of conveyance, had stayed to take off their meagre harvest—insufficient to meet the cost of sowing, insufficient to provide credit for another year. There they sat, silent, sustained only by each other's presence, hungry, without hope.

Gradually, gruffly, yet without undue intrusion, Mitchell drew from them: 'If we could only get to the siding. If we only had our fares to Perth, our parents could at least give us shelter and Bill could receive unemployment relief.'

Immediately that had been established Mitchell said: 'Get packed and I'll take you to the siding and while I'm unloading the car Mr Gabbedy will check to see if you have any clearing moneys due to you.'

I'd never been on the place before but it didn't take long to return and tell Mitchell there was £12 due (the limit that could be paid in any one month) and write out a cash order (the Agricultural Bank's equivalent to a cheque). The man looked amazed. He couldn't remember when or where and was sure there had been a mistake, his way. What should he say? His brain whirled; his pent up feelings broke when Mitchell put some notes in his coat top pocket.

'Shut up', growled Mitchell. 'That's not for you, it's for the kids.'

They left; their meagre possessions: what they stood up in and could carry with ease. I kept myself busy, making camp. That evening, over the second billy of tea, Mitchell, without preamble, remarked: 'Gabbedy, poverty is a dreadful thing, but despair is worse, worse even than death. In death you have oblivion, in despair you suffer the pains of all hell.' A pause, and a few puffs of his pipe, then: 'Have you got a spare quid you could let me have till I can cash a cheque?' The incident was closed.

1. F. K. Crowley, *Australia's Western Third* (London, 1960).

1a. A North American term; the Australian equivalent: short quarter = 500lbs, long quarter = 560lbs.

2. Western Australia's wheat sowings:

	1929-30	1930-31
Area sown	3,568,000 acres	3,955,763 acres
Yield	39,081,000 bushels	53,504,149 bushels
Average per acre	10.95 bushels	*13.53 bushels

*Paradoxically, the yield per acre had been exceeded only once before in the present century, 13.6 bushels in 1903-04.

3. *West Australian* 1 November 1930.

4. Unemployment figures for the period are incomplete and cannot be compared with current statistics. However, comparison of data from the 1933 and 1966 Censuses does indicate the relative magnitude of unemployment in the two periods. The 1933 Census data

showed that 13.6% of the work force was not at work; the 1966 Census showed that .6% of the work force was unemployed.

Trade union secretaries kept records of union membership and members who were unemployed. These indicate that unemployment probably reached its peak early in 1932. Union membership reached a peak of 27,689 in 1930 and then declined to 25,227 in 1932. In 1933 union membership represented 18.3% of all employees (including persons not at work).* It is likely that in the preceding four years a larger proportion of employees were union members and this sample would give a reasonable indicate of trends in unemployment. The 1933 Census data also indicate that about one-quarter of the persons not at work were members of unions.

UNEMPLOYMENT 1928 TO 1936—WESTERN AUSTRALIA

Year	Union Membership	Members Unemployed	Ratio %
1928	25,415	2,084	8.2
1929	26,386	2,633	9.9
1930	27,689	5,339	19.2
1931	26,537	7,249	27.3
1932	25,227	7,442	29.5
1933	25,278	6,257	24.8
1934	26,449	4,707	17.8
1935	27,708	3,710	13.4
1936	28,335	2,296	8.1

Source: *Commonwealth Year Books,* 1929 to 1937.

The above table shows that 7,442 union members were unemployed in 1932, almost three times the number in 1929. The trend in unemployment shown in this table is supported by an index of employment in factories. With the 1928-29 financial year as base, the index indicates a rapid decline in employment between 1929-30 and 1930-31. Employment reached a low point in August 1931 and remained at a low level till February 1932. In October 1932 the first real signs of improvement became apparent, but it was not until November 1935 that factory employment regained its pre-depression level.

* Persons who were (i) unable to secure employment; (ii) temporarily laid off from their jobs; and (iii) not actively seeking work at the time of the Census on account of sickness or accident, industrial dispute, resting between jobs or for any other reason.
(Material in this footnote was prepared by R. J. McKenna, Economist, R. and I. Bank.)

5. The salary cuts for the state Public Service comprised: Hospital Fund Tax (No. 39 of 1930), assented to 15 December 1930: a tax

of 2½% on all salaries; Salaries Tax (No. 40 of 1930) assented to 1 January 1931:

Rates varied:	£100 - £185	18%
	£185.10.0 - £250	20%
	£250.10.0 - £500	22½%

In line with the Premiers' Plan, accepted Australia-wide, the Salaries Tax was repealed and replaced by the Financial and Emergency Act (No. 20 of 1931), assented to on 18 August 1931. The act was allowed to lapse at the end of 1935 and cuts were restored. The Hospital Tax continued until the introduction of Uniform Taxation, which became effective as from 1 July 1942.

6. The Government of Western Australia, during the bleak days following the end of 1930, lived from week to week on what revenue they could obtain from all sources to pay wages and meet essential services. Every item of possible expenditure was examined by the Under Treasurer A. Berkley and three of his senior inspectors, G. L. Needham, A. C. Randall, and S. A. Taylor. So rigorous was their examination and enforced economies that the quartette were dubbed the 'Basher Gang'.
7. Western Australia, *Parliamentary Debates* 1930, vol. II, pp. 2255-9.
8. Summary of operations up to September 1935:

Number of Stay Orders issued	985
Number still in force	567
Number of satisfactory arrangements	218
Number of cases in which no agreement was reached	366
Applications refused	34

9. Between 29 May 1931 and 31 July 1931 the Commission sat on fifty days, of which evidence was taken on twenty-four days. Evidence was taken from: responsible bodies, 10; Agricultural Bank, 6; Merchants & Traders Association, 10; farmers, 13; Women's Section, Primary Producers Association and Country Women's Association, 1; British Medical Association, 1; country storekeepers, 6; general, 11.
10. The Farmers' Disabilities Commission, 'Report of the Royal Commission appointed to enquire into the disabilities affecting the agricultural industry of Western Australia 1931', *Votes and Proceedings of the Parliament of Western Australia,* 1930-31, vol. 3, p. 1150 *et seq.*
11. The Commission was constituted by the Commonwealth Grants Commission Act of 1933 and pursuant to the act, on 17 July 1933, the following members were appointed: the Honourable (later Sir) Frederic Eggleston, chairman, and Professor L. F. Giblin and J. W. Sandford.

In its initial stages the Commission considered compensation for disabilities arising from federation as a possible basis upon which to make its recommendation. It also considered the basis of financial aid. In its *Third Report* (1936) the Commission finally rejected the principle of compensation for disabilities arising from federation and chose instead the principle of financial aid, expressed in the following terms: 'Special Grants are justified when a State, through financial stress from any cause, is unable efficiently to discharge its functions as a member of the Federation, and should be determined by the amount of help found necessary to make it possible for that State by reasonable effort to function at a standard not appreciably below that of the other States.'

12. Also, in the *Government Gazette* issued on Friday 1 July, under the heading 'Classification of Agricultural Bank District Officers', appears the following:

1932—Titles of Senior Clerk	— Accountant
Paying and Receiving Officer	— Senior Clerk

Note this classification does not reduce the value of the positions as classified on July 1, 1931, the titles of the officers have been altered, positions created and others abolished. The completed decentralisation will involve the cancellation of positions in Head Office.

The above notification is very significant and from what could be gathered yesterday, in the absence of the Premier (Sir James Mitchell) who is also Minister in control of the Agricultural Bank, the Agricultural Bank Decentralisation Scheme originated outside Cabinet.

Indications are that the report of the recent Royal Commission on South West Settlement* was not followed in this re-organisation scheme, so far as the South West is concerned or possibly the Agricultural Bank decentralisation was decided on before the Commissions report was received.

Whatever the position in that regard the new departure will surely have far reaching effects on land settlement generally.

* Report of Royal Commission on Group Settlement 1925: chairman: C. W. Harper; members: W. D. Johnson, M.L.A.; C. J. Latham, M.L.A.; J. Lindsey, M.L.A.; the Hon. Thomas More, M.L.C.

15

The Royal Commission 1933-1934

Charles Mitchell was not a political animal. He was a practical operator who functioned most efficiently in the face of problems he could see and discuss and wade into; in the half-world of political manoeuvre he was like a bull in a china shop, and no event in his whole career so clearly demonstrated this as his appearance before the Royal Commission set up in 1933 to examine the affairs of the Agricultural Bank.

On 19 September of that year the newly appointed Lieutenant Governor, Sir James Mitchell, K.C.M.G.,[1] acting with the advice and consent of the Executive Council, named as Royal Commissioners Harry Hale of Perth, solicitor, Charles Diamond of Perth, banker, and Stephen Bede Donovan of Cunderdin, farmer. Their delegation was 'generally to inquire into, investigate, and report upon the transactions, affairs and operations of the Agricultural Bank of Western Australia'. It was paradoxical that Sir James, as Lieutenant Governor, should be required to assent to an investigation into aspects of agricultural credit and development on the policies of which he, as Minister for Lands and Agriculture and Premier, in all probability had the greatest single influence from 1909 to 1933.[2] It was he who, as Minister for Lands, in the 1920s, initiated the magnificently conceived (but badly executed) Group Settlement Scheme. His interest in the welfare of the small farmer, and his persistent encouragement of the class was the first positive move to develop the hitherto neglected South-West. It earned him the nickname of 'Moo Cow' Mitchell, curiously enough, since while he was in England in 1922 seeking financial assistance from the Overseas Settlement Committee for the extension of his Group Settlement Scheme to include English migrants,[3] he was dubbed by the English an Australian 'John Bull'. Sir James's simple, straightforward, unadorned presentation of his plan as a national ideal captured the imagination of the British public.[4]

While some investigation, some change, some modernization of policy was necessary and possibly long delayed, the searching inquiry mounted by the Royal Commission in the mid-depression years was ill-timed.

The Commission itself was well equipped: chairman Harry Hale, a member of a well-known legal firm, had the experience of the Farmers Disabilities Royal Commission behind him; Charles Diamond, then recently retired state manager of the Commercial Bank of Australasia Ltd; and Steve Donovan, retired wheat farmer and country storekeeper, had been chairman of the Land Board for the years 1928-30.

Nevertheless, the temper of the times was inimical to the coolest of judgements. People had been hurt: they were bewildered and confused—some to the point of despair. They were looking for someone to blame, prepared to lash out blindly, not knowing, nor caring, who might be hurt in the process. It is difficult to decide now whether the Commission afforded opportunity for a political advantage or a breathing space from criticism, or whether the state leaders were looking for a cure or a scapegoat; officers were castigated and subjected to severe public criticism—and there are many who felt that much of it was undeserved.

THE APPOINTMENT OF JAMES AS LIEUTENANT GOVERNOR in the previous July must inevitably have exerted, either overtly or covertly, considerable influence on Charles Mitchell. How would people now expect him to act? Must he not be more careful than ever of displaying unwonted superiority, or of laying himself open to the inevitable charge that he sheltered behind, or took advantage of, his eldest brother's elevation to one of the highest offices in the state?

Mitchell, however, was his own man. Now, more than ever, he would have been determined to stand on his own feet—to 'show them'. It is also consistent with his nature that he would have disagreed with James's authorizing the Commission so soon after assuming his office, but it is equally consistent with his code of behaviour that he never once mentioned his doubts.

His first contact with the Royal Commissioners raised his hackles. While being shown over the branch—their first official duty on arriving in Narrogin—they stopped to view a 'Proclamation' presented to him by the staff at the time of his brother's appointment to the Lieutenant Governorship, expressing their 'fealty' to Charles Douglas Mitchell, 'by the grace of Philip Collier—Premier, & accident of birth, now of Vice Royal lineage'. It was drawn on the back of the A.M.P. Society's calendar of the previous year, and embossed with bottle-top seals. When Mitchell proudly displayed it on the inside of his office door, one of the members exclaimed 'Childish!', tore it down and stuffed it in the waste-paper basket. It was not childish to Mitchell and, furthermore, it was his own property. He seethed inwardly and was certainly not in the right frame of mind for questioning on the following day.

Inspired by convictions and biased by family relationship and personality clashes, Mitchell immediately adopted a stand of—perhaps overdue—loyalty and support for the trustees; he certainly displayed resentment to certain lines of questioning and the consequent implications. This open resentment grated on the Commissioners and was certainly not lessened when, in reply to questioning, he conceded that while it would have been in the best interest of the Bank if he had shifted his field officers every three to five years, he refused to do so until they were either provided with official residences, or were protected against personal loss in the disposal of their own homes. He would accept no compromise—his staff's interests were at least as important as those of the 'trier' clients. Furthermore, Mitchell alone of the branch managers confessed he did not possess all the skills required by the job; and neither would he admit that any fault, at any time, was attributable to his staff—all decisions had been his and he alone was to be held responsible. Personality clashes occurred, and his deliberate manner provoked antagonisms, of which some took a long time to die down.

Steve Donovan, the only surviving Royal Commission member, relates how, feeling sorry for Mitchell because of the beating he took at the first day's questioning, he drew him aside that evening in the Hordern Hotel.

'May I have a few minutes of your time, Mr Mitchell?' he said.

'Alright, but call me Charlie.'

'Charlie, I suggest that tomorrow you bring your accountant, Mr Walters, with you, and refer any questions to him regarding office procedure on which you may not be clear or informed. As it is going at present you are only antagonizing Mr Hale unnecessarily.'

'Right', said Charlie. 'Thanks.'

No bad feeling remained with Donovan, whose admiration for Mitchell grew with extended association, and whom he described variously as: 'A Jack Blunt'; 'honest as the sun'; 'best outside man the Bank had'; and 'last of his generation'.

The Royal Commissioners reported at some length on their inspection of the Narrogin district of the Agricultural Bank—Charles Mitchell's territory—'as in their opinion the method of conducting the affairs of the Bank in this district furnished the clearest evidence of the urgent need for the re-organisation of the country branches and field work of the bank'.[5]

Of the branch management at Narrogin the Commission reported:

> The impression formed is that the branch should be managed by an officer with a sound knowledge of office routine and of the duties usually associated with a branch manager's position; the present branch manager is popular and possesses all the necessary qualifications for an outside man, he is conscious of the fact that his knowledge of office management is limited and his services could be utilised to greater advantage in supervising inspection work and control of the bank's securities. It is highly desirable that the control of the office be entirely vested in an officer fully conversant with branch management and office routine.

Moreover, the report stated the manager (Mitchell) had little knowledge of what might be required of a bank branch manager, and had been obliged to refer to his accountant (R. R. Walters) for information in respect to matters of ordinary routine.

> The Manager appears temperamentally unsuited for the worries and mental strain inseparable from the position of a bank branch manager. The accountant has a knowledge of the duties required of him and might be expected to have a wider knowledge of the branch accounts than that possessed by the manager, but in personality compares unfavourably with the branch manager.

In respect of one of the many alleged complaints the Commission acknowledged:

> The manager of the branch is away from the office for three weeks at a time and in the aggregate is away from the office between three and four months of the year. During those periods the management of the branch has been left to the accountant; correspondence relating to the business of the branch, addressed to the manager in his private capacity, is not opened by the accountant, and in consequence delay in answering such correspondence naturally follows, to the dissatisfaction of the bank's clients.

Narrogin was the largest branch. Its district extended from Darkan in the west to the Lakes areas on the east, and from north to south a distance of 60 miles.

Statistics published at the time of the Commission disclosed:

Area in square miles	13,500
Acreage	8,640,000
Number of settlers	1,627
Total of advances outstanding:	
Principal	£1,855,394
Interest	£398,274
Office and field staff	21
Total number of accounts	1,789

Narrogin ranked below Geraldton and Katanning in area, but was first in all other respects.

The Commission acknowledged that they could find no suggestion from Head Office or District Office that the administration was unwieldy and that any new branch office should be established.[6]

They were further aware that the field inspectors stationed through the length and breadth of the district were provided with neither quarters nor telephones. They had to find not only their own accommodation but also office and filing space.

'If these field inspectors wish to communicate by telephone with Narrogin, they must use a neighbour's or a public telephone, and Narrogin cannot communicate with them direct by telephone', the Commission stated. 'In the case of Lake Carmody, this would be impossible, but in every other case there is telephonic communication with the various centres, such as Kondinin, Lake Grace, etc.'

Particulars of administration costs were extracted for the information of the Commission in respect of the year 1932-33, covering Head Office and all branches. The average cost of administration per account at Narrogin branch was £4 4s 10d. Only Katanning branch was lower at £3 14s 3d. The overall average was £6 11s 4d.

The Commission reiterated an observation in the light of which one might be forgiven for wondering who had worn blinkers for so long.

The trustees of the Agricultural Bank of Western Australia consisted of the managing trustee (full time) and two co-trustees appointed from time to time for a period of two years and paid a sitting fee of £3 3s 0d, such payments not to exceed £250 per year. Referring to this circumstance, the Commission observed that when the bulk of the work thrust upon these trustees was considered, it would be realized that no government could expect efficient control, by one full-time and two part-time executive officers, of a bank handling £14,000,000. One wonders whether similar charitable thoughts might not have been expended on Mitchell, who at the time of the Royal Commission was receiving the princely salary of £475 less 22½ per cent as compensation for a full year's work, often of seven days a week and often from dawn to dark, in the control of nearly £2,000,000 of capital assets spread over 13,000 square miles among hundreds of difficult, demanding, different people. He suffered badly in comparison with his contemporary trading bank and stock firm managers both in salary and fringe benefits, but no mention of this was made by the Commission.

This item was one of the few upon which his friends can recall his having expressed dissatisfaction. His resentment certainly was not lessened by the publication of a further finding: 'The Commissioners are of the opinion that the management and control of the district managers and field staff have been very ineffective. The Managing Trustee admits he should have appointed more executive officers, including a Chief Inspector.'[7]

Political interference is a favourite catchword in bureaucratic circles and, as usual, it was cited as the reason for many of the Agricultural Bank's shortcomings. The Royal Commission in its findings stated:

> No. 5. The allegation of the Trustees that, in advancing recklessly they did so under Ministerial direction, has not been proved by the evidence submitted to your Commissioners, save in respect of the Esperance Land Settlement, the Murchison Miners' Settlement at Kalannie, and the 3,500 Farms Scheme which includes the Miners' Settlement at Southern Cross. All political interference in the management of the Bank and in the control of its policy must be abolished.

What the Commissioners neglected to emphasize, or perhaps did not even realize, is that any bank, to have a chance of success, must have discretionary powers in choosing its clients. This point was emphasized, however, in the evidence of Clarence George Russell, a farmer and soldier settler at Wickepin, representing the Primary Producers' Association.[8] He stated:

> It should be now quite obvious which settlers are not suited for farming, and they should be repatriated. I think that the Agricultural Bank has done, and is doing, more than any other financial institution to

> carry on and develop the farming industry. It has done more than its fair share, and should not be compelled by political or any other pressure to carry those who never did, nor ever will, make even moderately successful farmers—even under favourable conditions.

The managing trustee, A. E. McLarty, stated in his evidence that his department 'always had the land classified. The land was classified but not the men on it', and this untenable situation was complicated further by the fact that during the 1920 land-boom, wheatbelt blocks advertised by the Lands Department as available for selection carried the addendum 'Agricultural Bank Loan available'. Some authority other than the Bank, acting in good faith but without continuing responsibility, chose the client: the Agricultural Bank had no option but to accept the liability.

It is possible that political interference was never intended, but it would appear that the mandatory, indiscriminate granting of loans against farms of all sorts, in all areas, to all types of settlers, lay at the root of the trouble in which the Agricultural Bank found itself. Even so, there were varying degrees of justification, political and economic, for this policy. The country was young and to develop it was imperative; perhaps the risks were calculated, and today they could be accepted as worthwhile.

The Royal Commission records in its review of Land Settlement:

> Your Commissioners do not attempt to criticise the policy of settlement or the railway construction which has been carried out, but bearing in mind the want of uniformity in the soils of the agricultural lands in Western Australia, and the separation of the areas of good lands from each other by extensive tracts of worthless country, they cannot refrain from expressing regret that Bank advances and railway construction from the 'Nineties' onwards had not been co-ordinated, and the Agricultural Bank advances limited to those areas within a reasonable distance from the railways determined after consideration of the class of agricultural lands in those areas to be served.
>
> The acceleration of settlement, however, and the settlement of the country on a face, with good and bad lands mixed indiscriminately, seems to have been the sole object of the Trustees and Government, and we now have the resultant losses to be met.

Perhaps it would have been generous, and at least kindly, on the part of the Royal Commissioners then to have quoted the concluding sentences of the Honourable Sir John Forrest's speech when moving the second reading of the Agricultural Bank Bill on 5 September 1894:

> I take my stand on this Bill, and have no fear of the result, if its provisions are carefully administered. If a little loss does occur under it, which I do not anticipate, yet it will not have been thrown away, but will be spent on the lands of the colony, in a way than which no man can spend it better—in trying to subjugate the forces of nature, and in trying to improve this country, and make that produce something which is producing nothing at present. I cannot understand how the provisions of the Bill can work disadvantageously to the country itself; and, as I said before, it will not only be productive of great good but, if carefully administered and wisely carried out, it must prove an immense benefit to the people without causing any loss to the country.[9]

Who, or what really was to blame does not concern us now. One includes this brief mention of the Royal Commission of 1933 only because it is worth study in its application to the conditions and feelings—some of them very personal—of the time, and because it serves to highlight the yawning chasm which always, almost inevitably, separates the man on the job from those who view his labours, in almost dilettante ignorance of the day-to-day sweat, frustration, effort and achievement, from afar.

The Royal Commissioners, caught up as they were in the emotions and demands of the time, when the whole world of what had been so laboriously accomplished seemed to be crumbling about the community's ears, could not see the wood for the trees; and from our comforting eminence of hindsight we should not blame them for failing to realize what is now so clearly apparent: that in criticizing Charles Mitchell and his ilk for failing to do more than they did, they were laying their tongues against men pre-eminent in their own field, with no illusions of grandeur.

If Mitchell lacked anything, it was the support of the administration and an effective liaison with it, shortcomings which were applicable to the entire system. Head Office and branch staff were separate entities. Except at very minor levels one never moved from Head Office to the branches, and promotions to Head Office were the exception.

One limited form of liaison which did exist, however, was the district inspectors' conference, convened at infrequent intervals in Perth by the chief inspector, and at which discussions invariably revolved around extensions of development, and methods and standards to be employed. They cannot have been of very much use, nevertheless. Second in charge, Wally Cover, can never recall Mitchell having returned from one of these conferences mentally stimulated or enthused.

'Damned waste of time. Talk about everything except what really matters!' was his usual comment.

A suggestion from Cover to Mitchell in 1925 was rather prophetic. Cover was concerned over the practice of paying to farmers large credit balances of up to £2,000 on their clearance from the Industries Assistance Board, irrespective of advances secured under the Agricultural Bank Act.

'Why can't something be done about retaining these balances?' Cover wanted to know. 'They would not only be of help to the Bank, but we could make sure that the farmer used them wisely.'

The decision of the particular conference at which the matter was raised was that such action would cost more than it was worth. It took another twenty years for the thought to take shape.

The Commission's report was completed and signed on 25 May 1934. S. B. Donovan recalls his decision begging to disagree on certain aspects which led to his submitting a minority report. Donovan's disagreement with the main report lay primarily in item (iv) of clause 5 of the report, 'Future Policy of the Bank', which read:

(i) Consolidation of securities
(ii) Liquidation of securities
(iii) Adjustment or conditioning of debts
(iv) Discontinuance of advances

In his opinion, while the recommendations of the Commission sought to modernize the administration and to rationalize the debts of the Agricultural Bank's clients, to discontinue advances would have led to a stultification of the Bank's original purpose. The government's acceptance of this policy on the premises outlined would permit the Bank's competitors eventually to 'pick its bones', literally to take over the sound accounts at their convenience, leaving the 'skeleton', the weak accounts, for eventual liquidation, to the state's loss and the farmer's disadvantage.

Donovan had confidence in the future; he was not prepared to agree that the then reigning price of two shillings per bushel for wheat was likely to continue indefinitely; nor did he see it was consistent with the cost of production. Three shillings per bushel in the near future was the estimate he was prepared to work on, and which conditioned his thinking. He also believed that the bankers, merchants and storekeepers were as much in the farming business as the farmer, and that they should compromise with the man who did the work while they gambled on the annual return of the property. He spoke very caustically of the past laxity of the trustees, prompted by actual losses of £951,047 written off as at 30 June 1933, and possibly attributable to political interference:

> The Trustees must accept responsibility for much of this, and subsequent losses, particularly in respect to accrued interest as referred to elsewhere in the principal Report. This laxity has assisted in the development of a standard of business morality that threatens to undermine our social structure, viz., 'the utter disregard of an obligation to honour just debts.'

He went on further to enlarge a particularly pertinent point:

> The Agricultural Bank's constitution did not permit of the control of the wealth produced as a result of this development. Consequently, the business drifted through other channels; hence the Agricultural Bank, to meet administration costs and other commitments, had to rely mainly upon the difference between the rate of interest paid for its capital and the rate charged to its clients.

Donovan believed wholeheartedly that the Bank must be permitted to continue as a positive and progressive entity. The state was entitled not only to protect its own interests, but to foster them to the farmer's and its own future and mutual advantage. This ideal is summed up in the final paragraph of his minority report: 'It is important that the future credit of clients should be strictly controlled by the mortgagee to avoid past errors of the indiscriminate purchase of non-essentials.'

WHEN THE COMMISSION'S REPORT WAS PUBLISHED in May 1934 feeling ran high among Mitchell's friends because of the treatment he had received. Why was he, who had carried the heat and burden of the day for twenty-five years, who had met every challenge, implemented every change of policy with the minimum of assistance or advice, not lauded rather than condemned?

'C.D.M.'s' contemporaries felt that he had himself to blame—his nature revolted against passing the buck. Political pressures were anathema to him, and in his opinion the Royal Commission smacked of political pressure.

Furthermore, he considered that his boss, the managing trustee, A. E. McLarty, was being made the scapegoat. It seemed to him that the Royal Commission forgot how land settlement schemes had developed in the past, and appeared now to lay all the blame upon those who had battled with almost insufferable problems for years, within the bounds of their limited authority, and who also would now become responsible for clearing up the resultant situation—for writing off expenditure incurred long before, for creating a situation in which the primary producer would have a reasonable chance of making a success of his holding.

However, although his friends might simmer with indignation over his treatment by the Commission, Mitchell was not one to waste time in the irredeemable past. What he had done, he had done to the top of his bent and with only one goal in view—the benefit of the state through the well-being of the farmers who were then, and for some considerable time to come, its principal source of income.

With already a long, vigorous and successful career in the Bank's service behind him, and the deliberations of the Commission already rumbling away into the limbo usually reserved for the deliberations of Royal Commissions, he might have been excused for anticipating a further stretch of crowded years in the work, and among the people, he knew and loved.

Just around the corner, however, was one of the greatest challenges of his career, but not the advancement he might have dreamed of.

1. Sir James Mitchell's government was defeated in the April 1933 elections, with Mitchell also losing his seat to A. R. G. Hawke. His resignation as Premier and Treasurer was accepted on 24 April 1933.

   His successor was the Hon. Philip Collier, Labor Party leader who, in a most unprecedented gesture, nominated Sir James for the Lieutenant Governorship, to which he was appointed on 11 July 1933. He filled this high office with purpose and distinction—serving as Lieutenant Governor until 4 October 1948, when he was re-appointed as Governor, which position he retained until 30 June 1951.
2. Positions he held for a total of ten years and four months, exclusive of the three years and two months as Minister without Portfolio 1906-09.
3. The Empire Settlement Act of 1922 gave power to the Secretary of State to co-operate with the government of any part of His Majesty's Dominions in formulating and carrying out schemes for affording joint assistance to persons who intended to settle in any part of His Majesty's Overseas Dominions. The administration of the

Empire Settlement Committee, president of which was Rt Hon. L. S. Amery, M.P., Secretary of State for Dominion Affairs. It is worth recording that this was the first occasion on which the British government had been willing to make funds available for promoting emigration from the United Kingdom to the colonies and dominions.

4. The draft agreement to the extension of Sir James's 'dream' (the Group Settlement Scheme) was completed in 1922 between Britain, the Commonwealth government and Western Australia and signed in 1923.

   Western Australia agreed, at an estimated cost, excluding passages, of £6m, to settle about 75,000 new migrants (men, women, and children) from the United Kingdom and to establish about 6,000 men from such migrants on about 6,000 additional farms.

   Speed then became the essence of the contract. The blocks for the migrants to occupy were hardly being surveyed the day the first ship sailed from Tilbury. Only Sir James's optimism, drive, and authority quietened the doubters, and strengthened the believers.

5. *The Agricultural Bank Royal Commission Report,* 1933. All subsequent quotes from the Commission are from this source.
6. In January 1938 a branch office was opened at Lake Grace, relieving Narrogin of the eastern half of its district.
7. On 6 February 1933 Executive Council approved the creation of the position of Chief Inspector, Wheatbelt (Classification £495-£595). The Bank, in the following April, recommended R. D. Heuston, sub-manager, Perth office, with both field and administrative experience, for the position; but, on later consideration and in view of possible re-organization of the Bank, it was considered advisable to defer the appointment until future policy had been defined. Unfortunately, Heuston died in the following year.

   E. G. Kelso had occupied the position from 1920 to 1930 when in the last year of his service, he was transferred to district inspector, Katanning. This action may have been prompted by the compulsory absorption within the Bank's staff of permanent officers from the Group Settlement Department, one of whom, E. H. Faulkner, assumed the title of chief inspector, which he held until his appointment to district inspector in 1932.

8. Russell leased his property at Geetabin, with option of purchase, when he enlisted in World War I. On discharge, he discovered, to his dismay, that the lessee wished to exercise the option. Russell then selected the partly improved property 'Yarling Brook' at Yarling under the Soldiers' Settlement Scheme. He was unstinting in his praise of Mitchell to whose patience and encouragement he owed everything.
9. Western Australia, *Parliamentary Debates* 1894, vol. VI, pp. 480-92.

16

The Aftermath 1935-1937

The new Agricultural Bank Act (No. 45 of 1934), consolidated and modernized, was assented to on 5 January 1935 and proclaimed on 18 March. It was a compromise with the conflicting reports of the Royal Commission, the Commissioners' report and Donovan's minority report, and assured that the Bank would continue to function effectively. The reconstituted Bank would be serviced financially through parliamentary appropriations and borrowing by the Commissioners under the provisions of the act. But it still remained purely a mortgage institution: its future was limited. A further concrete result of the recommendations of the Royal Commission was the appointment, simultaneously with the proclamation, of three Commissioners: Hon. Alexander McCallum (chairman), Charles Lionel Clarke and Alexander Berkeley, the latter ex officio as Under Treasurer. The secretary of the Bank was the efficient, forceful secretary to the Royal Commission, R. W. Brownlie, who had assisted materially in the drawing up of the new act.

Alexander McCallum, M.L.A., was a dynamic character. He represented the seat of South Fremantle as a Labor member from 12 March 1921 until his resignation from the position of Minister for Public Works, Labour and Water Supply in the Collier government on 16 March 1935. His resignation to become a Commissioner was accepted as being in the interests of his party and the state, since he was taking over an onerous position, requiring positive leadership.

Having studied the report of the recent Royal Commission, he was intrigued not only by the challenge to revive the state finance institution designed solely to cover agriculture, but also by the minority report of S. B. Donovan, which reflected a generous insight into the difficulties both of the old administration and of the farmers, coupled with a serious view of the financial drift of the Bank and the farming community. The new chairman invited the experienced and down-to-earth Donovan to join the Bank in the position for which he appeared admirably suited—that of chief inspector who, in the terms of the recommendation of the Royal Commission, 'would devote his time to branch affairs, including the supervision of field officials'.

The Board pressed on to modernize management, and after diligent enquiries recommended the appointment of Cyril Abey as general manager. The latter joined the Bank on 1 September 1936 at a salary of £1,200, resigning from the English, Scottish and Australian Bank, Perth, where he had occupied the position of inspector of branches.

Immediately the new Board had taken over, all members of the staff were advised whether they were to be retained in the Bank or returned to the Public Service. C. D. Mitchell's appointment as branch manager, Narrogin, at a salary of £475, was confirmed as from 18 March 1935. Mitchell's reaction was noncommittal.

Disciplined by long years of loyal and arduous service, he looked forward to being part of the team of a revitalized organization bent on reform and consolidation. The details and technicalities of the new Bank Act did not concern him. These were for his staff to handle—his job was the protection and advancement of the farming community. The Bank's responsibilities were now one of consolidation, to save whom and what they could from the wreck. No new accounts were to be opened, and specifically the Bank's funds were no longer available to promote the traditional policy of extending land settlement, irrespective of changing economic conditions.

One of the Commissioners' first acts was to instruct Mitchell to report on the future of settlement in the Lakes district, a 70-mile strip of salmon-gum forest from 6 to 14 miles wide, stretching from Lake Carmody in the north to Lake King in the south (see Map 1); should settlement be continued and if so, in what form and at what level. The background of the instruction lay in the government's earlier decision to extend land settlement to the Emu Rock and Lake King districts on the understanding that the throwing open of blocks was contingent on railway facilities being provided subsequently by the passing through the areas of the lines to be financed by the Development and Migration Commission to service their proposed 3,500 farms scheme.

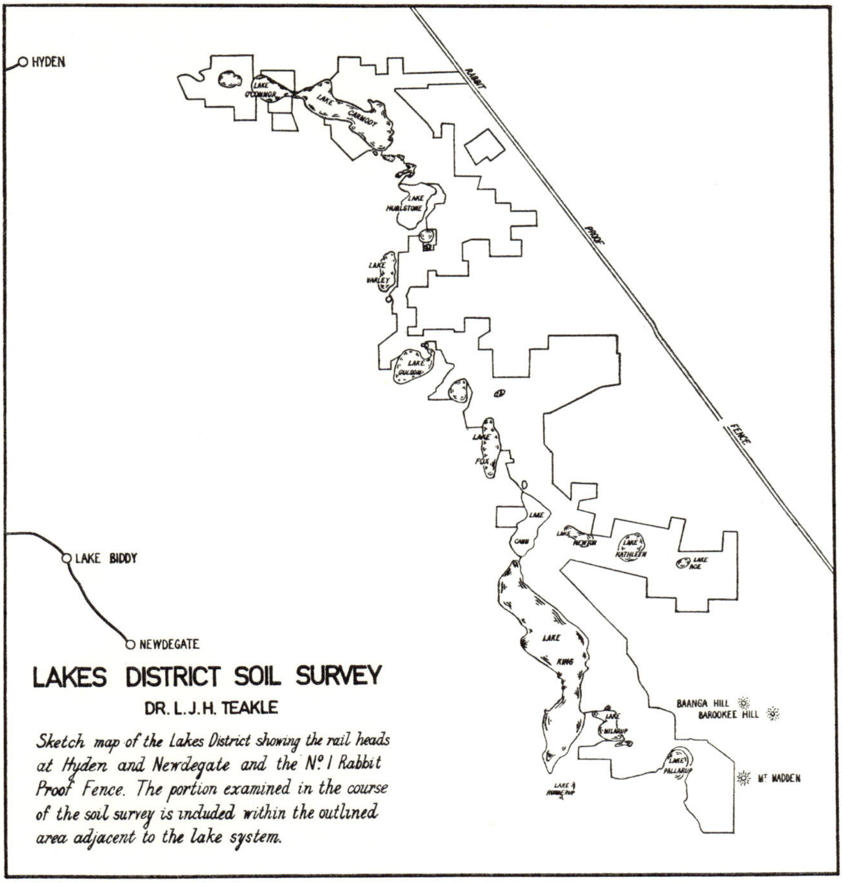

Map 1

The Commonwealth Development and Migration Commission had recommended, in principle, the establishment of at least 3,500 farms:

(*a*) south of Southern Cross and between Newdegate and Salmon Gums;
(*b*) north of the Lake Brown loop railway and east of the southern portion of the Wongan Hills-Mullewa railway.

The areas contained approximately eight million acres, seven million of which were in area (*a*). Not less than 1,000 of the farms were to be made available to migrants (see Map 2).

Under the terms of the Tripartite Agreement made between the British government, the Commonwealth of Australia, and

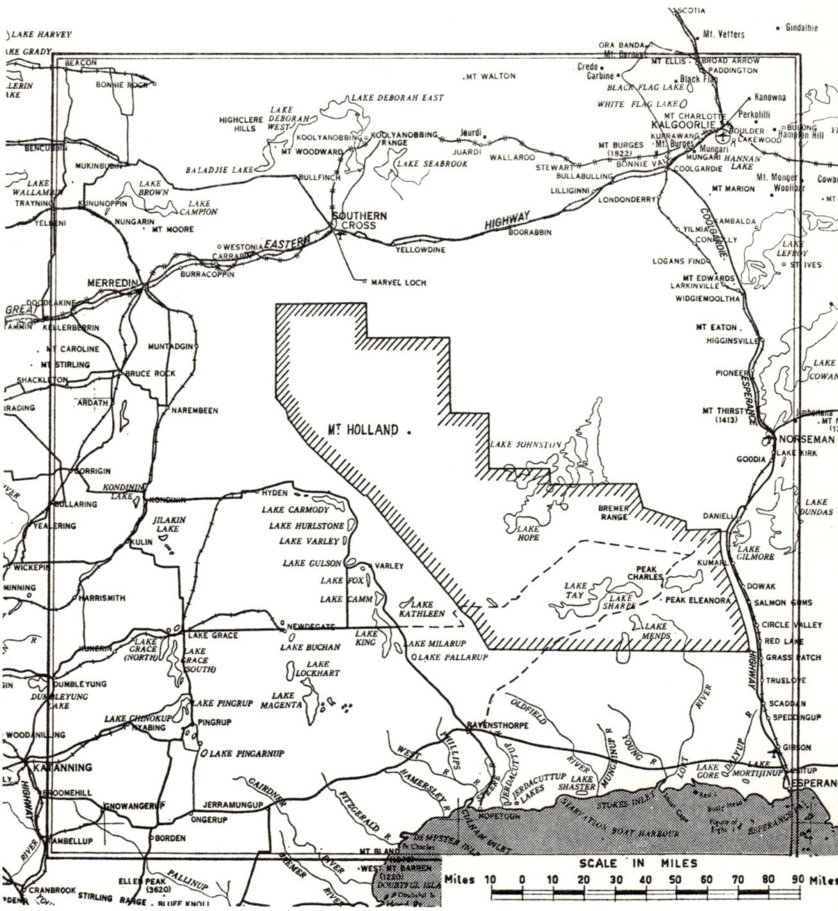

Map 2 Area examined by Dr Teakle as preliminary for settlement under the 3,500 farms scheme 1928-29

Western Australia, the state, on its part, agreed to absorb within ten years from April 1925 one assisted migrant for every £75 devoted to Public Works undertakings, while the United Kingdom agreed also that for every 10,000 assisted migrants, 3,000 should be members of families.[1]

In anticipation of acceptance by the imperial government, the Commonwealth government had provided £150,000 for water supply and roads in the areas to be opened up, and the state government, in 1928, confident of the adoption and success of this scheme, settled some 400 families in the Lakes, Gibb Rock and

Holleton areas and in the Miners' Settlement south of Southern Cross.

Hardly had this move been completed when the Development and Migration Commission, late in 1929, under direction, refused to proceed with the scheme. The onset of the Depression brought to an abrupt end the optimistic policy of populating the Empire's 'empty spaces'. The terms of the migration agreement with the United Kingdom were no longer in keeping with the ability of the state to absorb new citizens at the previous standard of living, and assisted immigration was immediately suspended.

To add to the general bewilderment came the publication of a report by the Agricultural Department's Dr L. J. H. Teakle, whose comprehensive soil survey of the Lakes district between August 1930 and May 1931 disclosed that the majority of the soils were too alkaline for successful wheat growing. The Agricultural Bank ceased advancing for improvements and the state was left with the liabilities, implied or fixed.

In the preliminary examination by Dr Teakle of the area recommended for settlement under the 3,500 farms scheme (see Map 2), it had been proven, at academic level, that the woodland soils in the areas south of Southern Cross were generally more saline than those of the then existing wheatbelt. In December 1929 Dr Teakle wrote: 'The investigations indicate that there is a very serious alkali (salinity) problem in the area in question. It must not, however, be regarded as a condemnation of the area. There is a close correlation between the texture of the soils and the incidence of alkali.' In accordance with the facts presented it was pointed out that the heavier soil types were most seriously affected and in an area of limited rainfall, the lighter soil types, which are relatively free of alkali, must be regarded as more suitable for wheat growing.[2]

As a consequence, it was accepted that this salinity problem would also affect the grain-growing potential of the Lakes district, and that future manifestations of salt would be more serious than in the older settled area.

Logically, perhaps, Dr Teakle was then commissioned to extend his research to the Lakes district and in August 1930 set out with a well-equipped field team to ascertain the extent of the problem. His soil survey covered an area of 228,667 acres, over a distance of 70 miles, where the annual rainfall ranged from 14 inches to less than 11 inches.[3]

Whatever the rights and wrongs of the reception of the report and the Bank's reaction to it, by the end of 1934 morale was at its lowest in the Lakes area. Of the 203 holdings selected and improvements initiated, there remained only 101 farmers under the Agricultural Bank and eight independent settlers. 'The soil survey was used by the Agricultural Bank as a basis for reconstruction' (*West Australian* 25 September 1935).

It was this area on which Mitchell was instructed to report: should settlement be continued, and if so, in what form, and at what level?

Here was an area Mitchell knew and believed in; he had designed the pattern of the original settlement and had maintained close contact with the settlers. The Lakes district, he maintained, was not finished, 'not by a long chalk!' All it wanted was a 'fair go'—he would prove it.

There were now two circumstances in his favour. Newdegate, 40 miles from the southern end of the settlement, was no longer the nearest rail head; under the 3,500 farms scheme, a section had been built from Lake Grace to Hyden, and opened in 1933. Also, he had the support of two outstanding young field officers whom he had personally trained and who were still in the area. They were F. M. (Frank) Collett, and A. M. (Allan) Martin, both of whom went on to bigger things. Each was promoted branch manager on 1 September 1935, and later transferred to the staff of the R. and I. Bank: there Frank Collett became chief valuer, and subsequently personal representative for the Minister for Lands in the Esperance district: and Allan Martin, after a period on post-war farm reconstruction and branch managerships, became the Bank's archivist.

Mitchell, assisted by his field staff, applied Dr Teakle's scientific findings to practical experience. They believed that a practical and economic answer lay in the linking of holdings, and in additional improvements which would permit eventually not only the carrying of 800 to 1,000 sheep, on 1,500 cleared acres (later increased to 2,000 acres), but also an adequate area of suitable soils for limited production of wheat. By now there was evidence that the problem soils might improve under a studied programme of pasture production and stock raising. Mitchell's subsequent report on the situation stressed the practical assessment of the district's needs, and covertly his own vision and logical approach to their solution, his feeling for the remaining settlers and his confidence in their ability to adapt, and to make progress with minimum assistance. The first constructive report in those bleak days of the Depression, this was a monumental piece of work, embodying exactly what farming communities needed: action which gave hope and advice which promoted confidence.

As was his nature, Mitchell generously acknowledged the assistance he received from A. E. Illingworth, senior clerk of the Narrogin branch, in the format and preparation of the report, the original concept and construction of which established a pattern later adopted by the Katanning branch manager, Walter Wardle, in his report on the reconstruction of the Nyabing-Pingrup areas and later when—as chief valuer—he reported on the Salmon Gums area.

Mitchell had a flair for pioneering and developing new concepts and ideas which led to preservation and support of settlement in most districts—meagrely perhaps on today's standard—but these areas remain today as a credit to their pioneers and the men who had faith in them.

ON 1 SEPTEMBER 1935 MITCHELL RECEIVED NOTIFICATION of promotion and a new appointment as senior branch manager 'for special work and relieving branch managers'. At the same time he was appointed manager of the Bank's Northam branch.

The news of his imminent departure had a numbing effect in the district he had served for so long. After twenty-seven years of active and fruitful association, the people of Narrogin and its districts had come to accept Mitchell as 'our property'. His public image was best exemplified in the attitude of the native population toward him. No man passed him by without touching his hat or forehead with a 'Good day, boss', or 'Good day, Mr Mitchell'. It was unthinkable that he would be with them no longer. Together they had weathered the worst days of the Depression, and they acknowledged Mitchell's courage and the confidence he instilled. The townspeople and farmers literally panicked: how much time had they to say goodbye? The *Narrogin Observer* reported on the news:

> Mr. Chas. Mitchell, District Manager of the I.A.B., has been appointed to Northam and expects to leave Narrogin at an early date. Charlie will be missed by a wide circle in which he has exercised considerable influence over the past 27 years; in fact, his departure may be regarded as a serious blow to the many sporting and other bodies with which he has been actively associated. Unless the numerous gaps

> recently made in the ranks of local citizenship are speedily filled it is certain that we who are left are in for a particularly hard time of it to keep various clubs and institutions functioning along safe and satisfactory lines without the help which they have hitherto received from these men. On present indications it appears very doubtful if this will be possible. However, those still remaining with the ship must endeavour to keep its sails trimmed to the wind until additional hands become available or the apprentices are sufficiently skilled to take the wheel. Meanwhile, good luck to Charles Mitchell in being piped down the side with musical honours, a cocked hat, and additional brass pips, accompanied by officer Alan Illingworth,[4] his strong and faithful henchman and a right good worker who also leaves to our great regret.

The following week the *Observer* recorded a send-off given him by the Race Club,[5] the first paragraph being particularly pertinent.

> Mr. C. D. Mitchell, president of the Race Club, was farewelled by members at a Dinner in the Hordern Hotel dining room on Thursday evening. It was a splendid tribute to his popularity that notwithstanding the fact the function had been arranged within a few days every member of the Club was present, including some who live in distant centres.

The fifty-nine members were his personal friends, but were also, to a great degree, representative of every sporting and public body with which he had been associated throughout the district.

It was rather a peculiar, and in many ways, an embarrassing predicament. He had been accorded an official send-off, had said goodbye to all his friends, severed his association with all Narrogin sporting and philanthropic organizations, but he had not left Narrogin. He felt his position keenly, doing his best to arrive unnoticed when he came home at least every second weekend, and hardly leaving the house until early the following Monday morning. He was meticulous in staying away from his old office and staff, partly because, without explicit instruction, he had no right to enter upon the preserves of another manager, and partly also because such visits rekindled old associations precious in the past but now discontinued.

Mitchell was never to take up his appointment as manager (elect) of the Northam branch. Practically coincident with his notice of transfer came instructions to report on the Bullfinch, Lake Deborah, Southern Cross (Miners' Settlement) and South Ghooli areas, and other 'special work' of like nature continued to keep him fully occupied. While Northam purported to be his headquarters, he actually worked from Head Office, and continued to live at Narrogin. There was no point in moving his family from one special area to another for the short periods during which he was engaged in them.

In their Annual Report for the year ending 1934 the trustees of the Agricultural Bank referred to the Miners' Settlement at Southern Cross:

> The grave position of this settlement was dealt with in the Industries Assistance Board report on its operations for the year ended 30th June, 1934. No active measure of land values exists at present on which an estimate of losses could be based, nevertheless, the rate at which arrears are accumulating is serious, but where the position, as in the majority of cases, is due to causes outside the settler's control, no advantage would be gained from forcing them off their holdings.

This particular investigation really concerned Mitchell. An intimate friend remembers receiving a letter from him at the time, stating that he had finished the survey and must now write his report. He had to make a decision which, it appeared, would be a distasteful one. In spite of the excellence of the soils, the districts would have to be judged on their history.

As far as wheat growing was concerned (and it was for this purpose that settlement had been designed), Mitchell condemned the area. He went further and spoke strongly against the standard practice of releasing fixed areas of Crown land for settlement irrespective of rainfall and suitability for type of agriculture envisaged; and of the inherent dangers of mono-agriculture. This was advanced thinking, some two decades before it was accepted authoritatively, and it is also one of the first official mentions of 'farm size' or 'economic units'.

The report was not well received, apparently because it tended to repudiate, quite bluntly, political beliefs of the day on agriculture and land settlement. Nevertheless, despite its cold reception, the report was adopted.[6]

Following the general pattern, the question arose in 1936 as to the practicability of overall continuance of the Ravensthorpe district as an economic farming area. Possibly because of its isolation, the paucity of records and to ensure an adequate examination, Inspector Wellard, of Narrogin district, was despatched to classify the soils on all Agricultural Bank holdings and to inspect and value the improvements.

Senior Branch Manager Mitchell was instructed to visit the same area after he had completed his work in the Bullfinch and Southern Cross areas, and 'to confine his enquiry to securities which are worthwhile carrying on, and that no action be taken by him to closely examine and report on those properties which can never be made use of for farming or grazing purposes'.

Mitchell arrived in Ravensthorpe in early August 1936, and for the first week was assisted by Katanning branch manager, A. M. Martin. The secretary of the Bank wired him on 31 August: 'Please advise approximate date when Ravensthorpe report will be completed.' Mitchell hated to be bustled, particularly when he worked with all expedition—and more so when the 'bustler' was one whom he disliked, as in this case. He made no answer, so a letter to the same effect was sent by the secretary on 3 September. There was still no answer, and a further telegram was sent on the seventh. Mitchell's reply was typical: 'Will be in Head Office tomorrow.' He took the report with him.

His opinion on the district is well summed up by his comments on the proposal for the linking of properties and his recommendations for reduction in price of land:[7]

> *Linking* I consider the majority of the settlers require more land to enable them to carry more sheep, but with few exceptions, they are not in a position to handle it. In most cases, I have not made any provision for linking as the general attitude of the settlers is one of hopelessness. There is ample land available for selection by anyone who is anxious to increase his holding and generally speaking, the land is cheap to clear—the majority only requiring to be rolled.[8]

A subsidized road transport scheme to Newdegate was recommended and later adopted.

Following the lines of Mitchell's proposals for the reconstruction of the Lakes district, the linking of abandoned Agricultural Bank properties generally in the lower rainfall areas, as a desirable principle, had proceeded. While it was acknowledged that further cleared areas would be required for most farms to carry additional stock, as proposed, it was even more important to proceed with the linking of the unoccupied holdings, in order to ensure the maintenance and preservation of existing assets and the control of regrowth. The preservation of assets did not refer solely to 'weathering and the vagaries of Mother Nature'. There existed a philosophy, peculiar to the times, that abandoned farms belonging to the Agricultural Bank were fair game for the removal of doors, windows, timber, galvanized iron, fencing wire and even whole buildings. The attitude could have been that in depleting one of the Bank's assets the 'thieves' were thereby increasing another, or that their need was greater than the government's loss; in any case, wasn't it a government of the people, 'whose job it was to look after the people—particularly farmers'.

Mitchell had received information that fencing wire was being removed from one of his abandoned farms, and as the inspector in charge of the area was in hospital, he sent a young officer from another district to investigate. Arriving in the area, the young inspector took the easy way of asking 'where' rather than of searching for survey pegs. He enquired his way from three men with a truck, apparently repairing a boundary fence.

'Where is so-and-so's place?' he asked.

'You go down this road for about 4 miles, then turn north. Take the left-hand fork about 2 miles up and you can't miss it', he was informed. The inspector was impressed with such courtesy and interest. Five hours later, having become hopelessly lost and redirected from the local store, 10 miles away, he arrived back to where he started from to find no truck, no men and no fence; but on the other hand, he had gained valuable experience and the local farmers had a story with which to regale each other.

Mitchell's time was fully occupied until the end of 1937. There were innumerable problems to investigate and there was always the sense of urgency, prompted by both political and farmer need.

No-one had come forward to pre-empt his leadership or authority in the particular fields in which he was engaged. A few, suffering from his forthrightness and his intolerance of bureaucracy, hoped for his replacement, but Mitchell had the confidence of the Honourable Alex McCallum, Chairman of Commissioners, another forthright person who detested vacillation.

McCallum had, through personal contact, obtained a high opinion of Mitchell's capacity as a practical agriculturist, stockman and 'camp cook', and had publicly expressed these opinions.

Mitchell 'liked the style' of Chairman McCallum and rendered him all deference and consideration. The late Frank Collett told the story of when the three of them stopped on the roadside for lunch near Muntadgin in early 1936. Mitchell had provided steak for the lunch and promised the chairman he would grill it 'like nothing you had ever tasted before'. Collett was rather surprised as Mitchell, poking around the boot of the car, suddenly requested: 'Collett, as soon as you light the fire, take Mr McCallum and walk him around this wodjil thicket to illustrate the association of this particular scrub with sheet laterite.'

The instruction was obeyed, but Collett was puzzled and, as soon as he could after lunch, whispered to Mitchell: 'What was the idea of getting rid of us?'

Said Mitchell: 'When I opened the boot I found the steak was blown; didn't matter about you, but the chairman had to be considered. That's why I wanted him away while I scraped it clean.'

Mitchell's deference toward McCallum was shown when he grilled the steak on a grid iron held on the end of a long green stick, as opposed to the usual procedure of flopping it on the coals.

Frank recalls that his stomach heaved several times during the afternoon when the chairman, on the other hand, literally drooling over memories of the meal, continued to compliment Mitchell: 'Either you are a good cook or you know how to choose your steak.'

In view of his seniority Mitchell expected to succeed to the chief inspectorship (wheatbelt) following the appointment of S. B. Donovan, the chief inspector (wheatbelt) to the arduous, exacting and thankless task of Chairman of Commissioners, subsequent to the sudden death of Hon. Alex McCallum on 12 July 1937.

The position of chief inspector, however, was never filled. While no official announcement was made it became apparent that the old order had changed and the functions of chief inspector were incorporated with the newly created position of chief valuer, to which W. Wardle had succeeded. Wardle was faced with the onerous task of carrying out a valuation, for the first time, of the Bank's assets, without evidence of market value of farming properties. There had been neither demand for, nor sales of, farm lands since 1929, and there were the allowances to be provided, against the rapid deterioration of former improvements on abandoned properties. It was only too patent that too many properties had been overcapitalized in terms of the farmer's ability to pay, but the capital liabilities, equalling actual cost, had been based on commodity prices ruling in the twenties, whilst earnings at this point of time were dependent on the depressed prices of the thirties. Nevertheless, Wardle did the job of the realistic revaluation of the Agricultural Bank's assets and adjustment of mortgage debts well, and this was an integral part of the revival and renewal of hope in the wheatbelt.

Political influences were reflected in two complementary items of legislation, one state, one federal, dealing with the individual farm debt structure. Were these planned with a time lapse of over six months between the events? Was it federal prompting and the promise of funds to assist all states, or did the state really intend to go it alone?

Section 65 of the new Agricultural Bank Act empowered the Commissioners to write off or reduce the aggregate indebtedness of borrowers on such accounts and for such sums as the Commissioners considered the securities might reasonably be expected to carry, or to adjust the repayments over whatever period the Commissioners might think fit.

There was a saving clause. The Commissioners were empowered to impose the condition, in any agreement involving writing off or reduction, that the other creditors, secured or unsecured, must enter into a mutually binding arrangement between themselves, the Commissioners and the borrower, for the reduction of their respective claims on such terms and conditions as the Commissioners considered reasonable.

Complementary to Section 65 and certainly most opportune, the Western Australian parliament, in the following year, 1935, were able to pass the enabling bill, the Rural Relief Fund Act, No. 17 of 1935.[9] The act purposed to place under the administration of an independent authority, three trustees, the expenditure of funds to be made available to the various states by the Commonwealth government under the recently passed Federal Loan (Farmers' Debts Adjustment) Act. Flowing on as an effective corollary, the trustees would also consider applications for advances to farmers who were under the operation of a stay order under the Farmers' Debts Adjustment Act 1930-34.

The Federal Loan Act authorized the Commonwealth government to raise £7m in order to make grants to the states for the purpose of enabling farmers to effect compositions with their creditors. Western Australia was expected to receive £1,300,000 from the first £5,000,000 received, and £260,000 from the balance of £2,000,000, making a total grant of £1,560,000 over three to four years. The state act was assented to on 12 November 1935. Trustees W. C. Angwin (chairman) and members H. K. Maley and W. A. White were appointed on 2 January 1936. W. A. White continued as director of both acts.

The new act had been given 'teeth' with considerable bite, and rightly brought about changes vitally necessary in any well-ordered economy, rectifying anomalies created under the 1930-34 Act. It now brought within its scope those who had not been able to effect a voluntary arrangement from their own resources, but who nevertheless deserved a similar opportunity. Some of the sections, dealing with the discretionary powers of the trustees and defining those farmers to whom no advance could be made, achieved a gleam of realism.

The Commissioners' first report on the operations of the Rural Relief Fund Act (1935) and Farmers' Debts Adjustment Act (1930) was contained in their Annual Report for the year ending June 1936, and is worthy of interest.[10]

Irrespective of the discretionary powers permitted the Bank under the Agricultural Bank Act, Section 65, and the Rural Relief Act, it would appear that the political philosophy of government during the 1930s, following the worst days of the Depression, was to reduce farmers' debts to a level which would not only restore morale and confidence (and, in the varying forms of compositions effected, give a meagre handout to their debtors) but would also provide for the servicing of the adjusted and secured debts at a level commensurate with the then ruling prices for farm produce. That commodity prices would ever recover so substantially and rise to the heights achieved in post-World War II years was never contemplated.

Under this philosophy millions of pounds were written off subsequently in adjusting the debts of Agricultural Bank clients, in the reconstruction of holdings and in the course of winding up Rural Relief Fund advances.

In the immediate post-war years farmers were advised that payment of 20 per cent of their outstanding F.D.A. indebtedness would cancel the balance of the advance. This offer they immediately accepted, encouraged by their banks to rid themselves so easily, at one stroke, not only of a shackling debt but also of the restriction which insisted that in regard to any property carrying a F.D.A. Act debt, no deal involving the land as security would be regarded as valid without the prior approval of the director.

There was a moral issue involved. When during the 1950s the state began to experience the first flush of the unprecedented affluence it still enjoys, a correspondingly brisk demand for agricultural land enabled farmers who had held on during the depression years to sell out for sums hitherto undreamed of; and few begrudged them their fortune, which had been hard won. Nevertheless, one is inclined to wonder how many of the newly rich then thought of the sums still morally due to the country storekeepers and their ilk. It is recorded that some did, but in a much smaller proportion than Saint Luke in his Gospel mentioned regarding the cure of the ten lepers, when only one of the ten came back to render thanks.

A school of thought, cautious and pragmatical, has developed (possibly from hindsight observation of the tremendous profits from sale of wool and farming real estate since 1950) which suggests that the farmer should regard his operation in the same light as that of the proprietor of any other form of business. If world commodity prices run against him or if he experiences prolonged adverse conditions, this is his risk. He should recognize it as such, and provide against it. When a farm fails, the new school of thought argues, why should the procedure not be exactly the same as for failure in any other form of private enterprise—with the farmer calling his creditors together to seek a composition?

Today, apart from subsidies and bounties, only a very small proportion of government money supports the wheat and sheep industries—on the state side there are some Delegated Agency funds, and on the Commonwealth side, the remains of War Service Land Settlement. No longer will it be accepted by the man in the street that tax-payers' money may be used to the same degree as in the now distant past, to bolster rural production with price support, or to aid farmers with grant moneys during periods of low production or low prices.

With consolidation, and experience gained by wheat and sheep farmers over the past four decades, and with the increased sophistication of governmental approach to, and provisions against, economic crises, it can never reasonably be contemplated that these sections of primary industry could again decline to the prolonged nadir of the 1930s.[11]

Neither is on-farm production any longer the hardcore of Australia's export trade, nor is it the major contribution to balance of payments equilibrium. Output from the non-farm sectors is increasing and strengthening. For the year 1969-70, and for the first time in Australian history, the value of non-farm exports exceeded the value of farm exports.

The productive value of farm land, once the farmers' guarantee of security, could be becoming his millstone. The individual farmer finds himself in an overcrowded industry, with a lessening importance in the total economy, and in danger of being phased out. His independence has been threatened and his status tarnished, and he could suffer further, as surely as his markets do. Each year there seems to be a decline, in total numbers and in proportion to total population, of people engaged in rural industries.

I suggest, however, that political economies would make it extremely difficult for an Australian government, influenced traditionally or otherwise, to ultimately resist withholding assistance, in some form or other, from the 'family' farmer, who may be affected again by prolonged adverse circumstances entirely outside any individual's planning or control.[12] Only his 'know-how', further informed by better education, and his inherent love of the land, can ensure full and safe use of the state's huge agricultural potential.

Then, if ever it should be found necessary to support any form of primary industry with debt adjustment from state or Commonwealth Treasury sources, *let there be no write-offs*. The debt adjustment should remain—interest free and repayment free, if this were considered advisable—a charge against the farming property for as long as it might remain in the possession of the grantee or his family succession.

Any such concessions should be considered morally and legally due for repayment. When recovered under any such scheme they would become part of a revolving fund directed to meeting those special demands which from time to time will become imperative in specific districts, and in specific rural pursuits, irrespective of conditions in the primary industries as a whole; but still as a recoverable charge.

Primary industry in the sectors of which I have written—wheat and sheep—will survive; decimated though it may be through economic attrition.

In suggesting any kind of assistance, one does not even consider its being directed toward the tax-saving, commercial-world absentee landlord, or to the 'out to make a fast buck' land-rapist.

The real farmer, however, tempered in the most recent crucible of experience, and with an upgrading of managerial and farming skills, will be encouraged to survive, and to create a way of life—a *family* way of life—a lasting asset not only for his state but for his children and his children's children. He will be encouraged to become efficient and stay in and to ignore the current dire prediction that he must 'get big or get out'.

It is apposite to record here, in the midst of Mitchell's depression troubles, that although in other districts throughout those unhappy years there were reports of shootings and threats of violence, of blacklisting and public protests following eviction notices, little or none of this occurred in the area under Mitchell's administration. In this respect there are numerous stories of Mitchell and the team he had assembled and it is well worthwhile to relate some of them. Inspector P. W. (Peter) Doig, who was transferred from Northam district to Mitchell's staff in 1930, and stationed at Lake Grace, says of Mitchell: 'He was the whitest man I ever knew, and one who never placed himself under an obligation to anybody. He was meticulous in making and breaking camp. Brushes were cleared away and neatly piled, papers burnt and rubbish buried. When he left the place was spotless.' Another memory of Doig's is of Mitchell's consideration for his staff and his forthright independence. The pair were together in early November 1934, carrying out a survey of abandoned farms in the Lake Biddy district. After breakfast, one morning, Mitchell said: 'Doig, I think we'll work our way into Lake Grace today in time to listen to the Melbourne Cup.'

The hotel had the best wireless, and there they gathered with the locals to hear Peter Pan win the second of his two Cups. Horses then became the topic and as everybody knew everybody, a good time was had by all—Mitchell finally electing to stay the night at the hotel rather than putting Doig to the inconvenience of accommodating him in his Lake Grace home.

At 5 a.m. next morning Peter, asleep on his side verandah, was aroused by Mitchell: 'Come on, Doig, let's get out of here.'

'What's the matter, Mr Mitchell?' he demanded.

'Some lousy so-and-so stole my wallet from under the pillow', Mitchell complained. 'Serves me right for staying at a pub.'

'What about breakfast?'

'Not on your life. They can't be bloody-well trusted with anything—I've got some bacon and kangaroo and two bloody sharp knives', said Mitchell.[13] On conclusion of breakfast, he suggested that as they were in the locality they should have a look at the property of an unsatisfactory settler. 'Doig, I think you've been too kind to him too long', Mitchell remarked. Accompanied by Doig and the settler, Mitchell inspected the farm and when he had finished he said to the settler: 'I don't think you're trying. You've had it. Doig will serve your papers, which he's been hanging on to for too long.' He then stalked off to the car.

'Sorry, M——', said Doig, 'but I've got a direct order now.'

'Best news you could give me', the settler replied. 'I've been fed up with the place for the last two years. I don't mind being kicked off, but I wasn't going to have it said I walked off!'

As an indication of the way in which Mitchell's precepts were inculcated in the men he had trained, I recall an instance which concerned the long-delayed eviction of a settler in another district. There was a large and young family, and there was nowhere for them to go, and less for them to do. With complete cessation of credit the settler could not crop, and grazed sheep only. They served as rations for the family, but the wool was secured under the Agricultural Bank's Statutory Lien. Any ready money came mainly from the sale of milk in the nearby town, from an increasing herd of cows to which the Bank's representative shut his eyes.

An opportunity presented itself when the Bank inspector met the settler in the town.

'There's a house on the edge of the town vacant', he said. 'I asked it to be held for you. The rent's eight bob a week.'

'Thanks for letting us stay as long as you have', the settler said, 'but how are we going to live?'

'Just make sure your cows are not on the property when I come out with the Distraint Notices', the inspector advised, and then added, 'And by the way, there's a couple of coils of netting in my shed you can have. They'll make a house paddock for the cows and they can graze on the commonage. Let me know when it suits you for me to come out.'

A few days later the settler called at the inspector's house. 'Could you come out on Tuesday afternoon?' he asked. 'We'll have afternoon tea ready.'

So Tuesday it was, and over a special afternoon tea, the settler's family acknowledged their eviction, the husband agreeing to become unpaid caretaker and his own bailiff till the clearing sale. That same family, now free and independent, made the most of their new opportunity; they never looked back and certainly had no regrets.

Despite some light moments, this chapter may have been found depressing since it continues to recount the Depression, its effects, its influences and solutions attempted. However, at this point of time in Mitchell's history, the end of 1937, progress could be reported. On the office side, the amalgamation of accounts, a most desirable measure, as provided for under the new act, was proceeding apace. Advances to farmers in the past had been over extended periods for gradual development of farms, with a separate ledger sheet required for each loan. Debts rested in both the Agricultural Bank and Soldiers Settlement or I.A.B. sections and I can recall individual accounts with over twenty ledger pages. Bookkeeping was complicated, involving much clerical work. Now, office work on ledgers was being halved and at last the farmer could understand the Bank's accounting system and their own personal statement of account.

The general outlook was one of hope: if there was not brightness, there was certainly less gloom. The state was slowly emerging from the trough of depression. Each year saw fewer men on sustenance. The basic wage for the Metropolitan Area had risen to £4 1s 1d, not yet up to the £4 7s 0d of 1929, but a substantial increase on the £3 8s 0d nadir of dismal 1933.

The 1930 Salaries Tax was allowed to lapse at the end of 1935. Salary cuts were restored, and the State Civil Service and Teachers' Union accepted a reclassification of salaries, as at 1 July 1936, with a resultant all round reduction of only 6 per cent on 1930 rates.

The area sown to wheat for 1934-35 totalled 2,764,000 acres which, with an average of 9.8 bushels per acre, yielded 27m bushels with return of 3s 2½d f.o.b.

The slightly improved 1937-38 sowing of 3,026,000 acres, with a 12-bushel average, yielded 36,225,000 bushels, with the heartening price of 5s 7d f.o.b. While the harvest had increased by 30 per cent, the gross value showed an improvement of 46 per cent.

Sheep numbers in the South-West Land Division had risen from 3,313,500 to 3,674,250 between 1934 and 1938. The price of wool in 1934, 15.65d, had near doubled that of the previous year, 8.57d. Despite the drop to 9.65d in 1935 the improvement resumed and steadied at 13.59d in 1936 and 16.16d in 1937.

The state Treasury 'rubbed its hands', the farmer found some comfort from having stayed on, willingly or otherwise, but the financial institutions were chary and wary. There was no spending spree.

Mitchell's hopes for a senior administrative position were dispelled when he was appointed manager of the Geraldton branch from 4 January 1938, with the sop of a salary increase of £9 per annum. He had given the lead and a great deal of original thought in the reconstruction of the wheatbelt and, although this work was finished for the time, many of his friends considered that his new appointment smacked of relegation.

The ostensible reason was that the position of branch manager of Northam had been filled by G. S. Gillies since early in 1936 and a vacancy existed at Geraldton through the transfer of Alex Whiteford to the newly opened branch at Lake Grace.

A place had to be found for Mitchell and the appointment could well have suited Head Office's hierarchy: Mitchell's forthrightness, practical experience and forceful manner were too strong a meat for them to stomach.

He would now be 434 road miles from Narrogin and would not be able, as he had been for the past year or two, to visit it every other weekend. The severance was physical, and it hurt.

The following chapter deals with the indelible impact he made on the development and character of the town he was now to leave, with what I consider to be his greatest single civic contribution, of which much is still extant.

It is a fitting introduction to the climax of his working life and his retirement, which followed.

1. In 1925 the federal and state governments signed a £34m agreement with the British government to provide loan money for development purposes in Australia, a portion of which Western Australia would receive. Western Australia endorsed the agreement on 19 October 1925.

   The federal government introduced the Development and Migration Act (No. 29 of 1926) which was assented to on 21 July 1926 and came into force on 1 October 1926. The Commissioners appointed under this act were: H. W. Gipp, chairman; C. S. Nathan, vice chairman; J. Gunn and E. P. Fleming, members. Their purpose was to exercise a general supervision over the different programmes of immigration and land settlement.

   The Commission's *First Annual Report* for the period ending 30 June 1927 reported regarding Western Australia: 'In Western Australia in addition to special attention directed towards the gold mining industry the Commission's investigations and collaborations with the State Government will, it is hoped, lead to a comprehensive program of balanced development over a period of five years or more. Such forward policy would include railway construction, land settlement, roads and water supply.'

The Commission's *Second Annual Report* for the following year ending 30 June 1928, recorded that the British and Commonwealth governments had agreed to the sum of £150,000 being provided for preliminary survey of all aspects of the 3,500 farms scheme.

The Development and Migration Commission was disbanded by the federal government in 1930. Its purpose had been served. Nearly two decades passed before U.K. migration was revived, but never again on a land settlement basis.

2. 'The soils of the 3,500 Farms Scheme area Western Australia', *Journal of the Department of Agriculture,* June 1939, p. 202 *et seq.*
3. J. H. Teakle, B. L. Southern and S. J. Stokes, 'A Soil Survey of the Lakes District of Western Australia', *Journal of the Department of Agriculture,* September 1940, p. 251 *et seq.,* stated, *inter alia*: 'The chief potential agricultural soils of the district were those of the salmon gum woodland (32,595 acres), the gimlet thickets (6,696 acres) and the better class mallee scrub (37,533 acres) in which salinity is not high. These soils, covering 76,824 acres in the area examined, resemble the soils of other parts of the wheat belt under similar rainfall conditions.

   'Problem soils were those of the morrel woodland (powdery or snuffy when dry) (59,700 acres) and areas of other groups where excessive salinity occurs (28,269 acres). Other soils of low agricultural value were those of the heath and wodjil plains (36,521 acres) and the low-lying, lakey country (18,323 acres). Small lakes and rock outcrops accounted for 9,030 acres. The woodland and Mallee scrub soils generally were rich in potash, lime and magnesia, normally low in nitrogen and humus and extremely deficient in phosphate.

   'Apart from the morrel woodland and saline areas it was considered the soils were suitable for general mixed farming, with stock raising and wheat production as the chief activities. Transport difficulties, however, were serious and a limiting cost-factor. Over-stocking and over-cultivation would need to be carefully avoided to prevent soil erosion and land deterioration.'
4. Illingworth was appointed accountant at Salmon Gums.
5. Details of the presentation and laudatory references are recorded in the chapter 'Horses and Polo'.
6. Of the original number of settlers totalling 334 (Bullfinch 223, Southern Cross 111), only 129 were in occupation at 30 June 1936, when they were returned to Lands Department control, free of encumbrance except for the assessed present-day value of stock and

plant, under bill of sale to the Bank, and for which the settler gave a hire purchase agreement. All debts to the Agricultural Bank, including interest, had been written off.

7. Price of land was to be adjusted from an average price of 10s per acre to: first class 5s, second class 2s, and third class 1s. It was recorded that the average rainfall over the growing period for the past thirty-three years was 1,012 points. These prices were later further reduced on the initiative of the Lands Department.
8. File records held by R. and I. Bank.
9. The relevant sections are recited hereunder:

   The Fund shall consist of any moneys provided by the Commonwealth for the purpose of rural relief.

   The Fund shall be under the control of three trustees who shall be appointed by the Governor. One at least of such trustees shall be a farmer.

   It shall be the duty of the trustees to consider applications for advances from the Fund to farmers who are under the operation of a stay order issued under this Act or under the Farmers' Debts Adjustment Act, 1930-1934, and who desire to utilise such advances in order to make compositions or schemes of arrangement with their creditors.

   Subject only to the provisions of Section . . . the trustees had unfettered discretion in regard to the advancing of funds to farmers for the purpose of this Act, but no advance could be made to any farmer—

   (*a*) who had not, in the opinion of the trustees, a reasonable prospect of carrying on his farming operations successfully if assisted under this Act;
   (*b*) who had shown by his past conduct in regard to farming operations that he was undeserving of such assistance;
   (*c*) unless, in the opinion of the trustees, some discharge of the debt or debts in respect of which such advance was made was necessary to ensure that the farmer would continue to carry on farming operations, and to give him a reasonable prospect of carrying on those operations successfully;
   (*d*) for the payment of a debt or a composition on any debt due to the Crown;
   (*e*) for the payment of a debt, or a composition on a debt, which is barred by any Statute of Limitation.

10. (*a*) Rural Relief Fund Act—In the adjustment of clients' indebtedness under this Act, the Commissioners have worked in harmony with the Trustees appointed under the Rural Relief Fund Act and their executive.

    While the compositions effected have necessitated writing down of the Bank's assets, the freeing of loose assets from outside creditors has been of benefit to its clients.

Applications for debt adjustment are dealt with under two headings: viz., Rural Relief Fund Act, Section 11, and Agricultural Bank Act, Section 65. Regarding the former, 1,089 requests for valuation of properties have been received to date, of which 850 have been fixed as far as the Bank is concerned, but a number of these have yet to be finalised with the Director of Farmers' Debts Adjustment Act.

Under Section 11, up to 30 June last, meetings had been held in the case of 186 mortgagors and proposals accepted by creditors. Of this number, the Bank has written down 61 accounts by £96,045 and on the balance, has amended the terms by capitalising interest arrears and extending terms of repayment.

In respect of the 61 accounts upon which the Bank proposes to write down a total amount of £96,045, averaging £1,574 per account, the proposals provide for an advance of £13,997 from Rural Relief Funds (Commonwealth Funds) to other creditors averaging £229 per account.

Valuations are now being requested by the Director at the rate of approximately 200 per month, and these necessitate special and urgent action, as received.

Regarding applications for debt adjustment under Section 65 of the Agricultural Bank Act, 1934, 348 settlers have applied and the staff is concentrating on the preparation of the balance of these assessments with a view to early finality.

(*b*) Farmers' Debts Adjustment Act—Under Section 5 of the Farmers' Debts Adjustment Act, 276 Agricultural Bank mortgagors were operating during the year, representing a total debt at 31st December, 1935, of principal and interest, £573,712. The number of accounts being carried on for the ensuing season is 232. All Agricultural Bank clients under Section 5 have been advised to apply for adjustment of their debts under the Rural Relief Act, as this enables coordinated adjustment of clients' indebtedness to the Bank and outside creditors.

The relief granted to any farmer shall be in the form of an advance, repayable by instalments over a period of not less than twenty years, but the payments shall not commence earlier than three years from the making of the advance.

Provided that the farmer may at his option repay the advance or any instalment thereof at any time before it falls due and notwithstanding that the said period of three years has not elapsed; but this section shall not be deemed to confer any priority over any other encumbrances existing at the date of the stay order.

It shall not be necessary to register any mortgage or security taken pursuant to this Act under Transfer of Land Act, 1893, or the Deeds Registration Ordinance, 1856 (19 Vic. No. 14), or any other enactment relating to the registration of land titles or dealings in land.

The machinery of F.D.A. Act was used for arrangement of meetings and compositions but all decisions reserved to the Trustees.

11. *Why it cannot happen again*:

The fact that an economic depression occurred in Australia can be attributed to economic phenomena. That it was so severe and lasted so long was due largely to an almost complete lack of expertise in dealing with economic fluctuations, with the result that any patched up measures taken in the hope of correcting the situation merely worsened it.

It is unlikely that the economic phenomena have changed, but our ability to measure and understand them has improved. However, there still exist conflicts between the policies put forward by different schools of economists. Thus in the United States, which for almost two years has been experiencing inflationary recession—a new phenomena but possibly one that could be as serious in its effects as a depression—conflicting policies have been expounded by the various policy agencies and academics. For convenience these can be divided into two groups: the new-Keynesians who favour fiscal measures and the Chicago school who favour monetary control.

What changes have occurred that make it so unlikely that inappropriate corrective measures would be applied now or in the future?

1. In Australia, development of official statistical collections and research organizations have kept abreast of developments in economic research, though we have been slow to make full use of some techniques such as input-output tables and linear programming. These developments mean that our policy makers are much better informed on current conditions than were their counterparts forty years ago. Today they have available timely measures of the level of employment and unemployment, of Gross National Product, of changes in the liquidity structure of the economy and so on. The greater sophistication of the economy has made it essential that this information, and possibly even more sophisticated data, be available. However, economic decision makers today are better informed, relative to the requirements of the environment, than were the politicians and economists of the early 1930s.

2. Institutional changes have also been important in strengthening Australia's economic position. Today we have a strong central bank occupying a prominent position in administration of economic policies. With effective control of taxation and better opportunities to raise loan money, the Federal Treasury is now in a position to instigate effective fiscal policies.

Shooting trips to Tarwonga and Culbin
Charles Mitchell centre

First Narrogin Sports Ground Board, 1920
Front row, W. F. Bird, J. J. James, P. M. Sullivan;
back row, A. E. Hardie, F. Tuohy, C. D. Mitchell

Memorial gates, Narrogin Greater Sports Ground
Charles Mitchell performing the official opening of the memorial gates, 1953

3. The structure of the Australian economy has changed and is continuing to change, so that we no longer depend on exports of rural commodities to keep the economy viable. This enables a wider choice of corrective measures and lessens the impact of changes in the demand position for these commodities.

4. Wrong policies were adopted in the early 1930s because people in responsible positions were unable to appreciate the key role played by investment and the subsequent problem of under-investment. Economic knowledge has improved to a degree where the role of variables such as investment and consumption are now understood and trusted. With better understanding of the underlying factors policy makers are less likely to make wrong decisions.

5. The fact that the gold standard has been abandoned (1933), together with recent improvements in international liquidity arrangements, has allowed for continuous heavy rise in the volume of international trade and enabled the elimination of the nexus between international parity rates and domestic economic stability.

12. On 19 August 1970 the Minister for Primary Industry Mr Anthony explained in the House of Representatives an emergency aid plan for woolgrowers and also that the Bureau of Agricultural Economics had been asked to investigate and report to him the immediate and longer term needs for debt reconstruction and farm adjustment (*West Australian* 20 August 1970, p. 2; and *Australian Financial Review* 28 August 1970).

13. A very necessary adjunct, together with good teeth, for eating kangaroo steak, fried or grilled.

17

Narrogin Club and Sports Ground

The Narrogin Club, still a vital influence in the life of Mitchell's beloved 'home-town', was formed just before the outbreak of World War I—mainly because a number of leading citizens liked each others' company and wanted to preserve the happy and compatible association they enjoyed. Anybody who has not lived in a small Western Australian country town before the advent of all the isolation smashers of today—radio, television, movies, milk-bars, proliferating motor cars, light 'planes, and so on—cannot even begin to imagine the degree to which the social and sporting life of such a community could hang on the personalities and activities of a very small group of its citizens. In Narrogin—no exception to the rule in those far-off and now almost incredible days—the live sparks were Mitchell and his polo-playing and racing familiars. It was they—A. C. N. Olden, Ted Hardie, Mitchell, W. L. Johns, George Lavater, Marshall Fox, Dr J. B. Lewis and the solicitor, D. C. Scott—who attended a meeting at Olden's home in either January or February, 1913, to decide upon the formation of the Narrogin Club.

Having pronounced the proposition practicable, the founders proceeded to obtain the requisite number of members to satisfy the Licensing Court and apply for registration. Apparently there was some opposition to the club's formation: an anonymous letter appeared in the *Great Southern Leader* (26 April), suggesting the proposed new club for Narrogin might not really be wanted, mainly because the existing facilities for beer drinking were, by some, considered to be adequate. The editor merely noted: 'During the week negotiations for lease of premises have been in progress.' Until they obtained the club premises, some fifteen months later, meetings were held in a shed at the rear of Harold Caldwell Ebsary's Agency, in the location occupied today by the Totalisator Agency Board, in Egerton Street.

Progress was slow, but sure. On 26 June 1914 the club, with premises situated at No. 12 Federal Street,[1] was incorporated under the Associations Act with George Geoffrey Lavater, secretary, and D. C. Scott, solicitor. The committee named in the memorial were Marshall Fox (president), Thomas Anderson Hardie, Arthur Charles Niquet Olden, James Brook Lewis, David William Hartnell Mackie, Nicholas Rogerson Anderson, Henry Ferguson Denny and Harold Lyle Facy. Mitchell begged to be excused; he was too busy and of conscience considered the committee members appointed more able than he. Application was then lodged for registration of the club under the Licensing Act, and certificate was issued on 4 December 1914. R. (Bob) Nairn was appointed steward. Tom and Ted Hardie were the trustees, and probably the club's guarantors at the local bank. The club operated at 112 Federal Street until 1927, when it purchased the present Fortune Street site from the Masonic Lodge. Although, naturally, members were selected, the club was never at any time termed exclusive. With generally a common interest of the Agricultural Society and Race Club, members desired a certain privacy in their relaxation, and from their association and relaxed discussions there flowed the spirit of community effort and progress.

Mitchell enjoyed the Narrogin Club, the atmosphere, the amenities; cards, billiards and snooker induced the relaxation he usually found so hard to come by. He enjoyed the friendships, the leg pulling, the repartee. Dr B. C. Cohen recalls the consternation caused at the club's Annual Meeting in 1920 when Mitchell challenged the accuracy of the balance sheet, claiming the bar stocks had been understated. The president, H. C. Baird, looked to the phlegmatic secretary, George Lavater, for guidance. George was so dumbfounded that he stopped puffing his pipe and could only shake his head. Baird broke the silence: 'Mr Mitchell, I hope you know what you're talking about.' 'Too right I do', replied the imperturbable Mitchell. 'What about the two cases of whisky Jackie James and I pinched last Saturday night and hid in the stables.' The stables were those provided at the rear of the premises for the convenience of members as horses still far outnumbered motor cars.

The meeting collapsed in laughter.

Allied with the fun were the challenges of solo whist and poker. 'Never for high stakes, and never with strangers'—were axioms Mitchell never departed from. 'Threepenny blind solo' and table limits of never more than ten shillings for poker restricted both losses and winnings, but nothing detracted from the thrill of winning from one's closest friend after a night of cut and thrust and no quarter given. Sometimes the sessions lasted till daybreak and Bricky recalls seeing his father moving around the house at daylight.

'What are you doing up so early Dad?'

'Feeding the horses.'

'What, with your best suit on?'

Next to cards and yarning, Mitchell loved a whisky. Never able to afford as much as he may have liked, he welcomed the introduction of Australian whiskies and particularly the 'Corio Special' marketed in the early 1930s.[2] It was to his taste and suited his pocket. He drank it for preference, whenever possible, and never neglected an opportunity to extol its virtues—it was a favourite discussion point. The distributors of this product missed an excellent piece of public relations in not including Mitchell's photo and unsolicited testimonial in their advertisements.

Clarrie Russell deprecated the Australian product, classifying it as 'blackboy gum polish'. Mitchell retorted that the only thing he could find wrong was 'there were only nine nips to the bottle'. Nevertheless, Mitchell neither drank more than he could afford, nor played cards for stakes beyond his means. As with all things he enjoyed, he gave to them a full measure of personal zest and just the amount of time he could spare to enjoy or further them. The children looked forward to one of Dad's nights out; they knew that for every so many of the rounds of drinks the winner of the jackpot paid for, there would be a drink foregone in favour of a bag of sweets for the kids.

No-one recognized more than Mitchell the danger of alcoholism, and there are numerous stories of his dealings with friends who, he thought, might be succumbing to drink. To these he tactfully made the suggestion, having manoeuvred the 'victim' into the right frame of mind and company: 'I'll bet you I can stay on the wagon longer than you can. Let's have a bet, say, for the next two months.' He never cheated and he had many successes. This form of abstinence was often self-imposed, and religiously adhered to, thoughout his life. But he would never accept a challenge to forego his beloved pipe, even for a day—he always said he knew his limitations—until old age and illness forced him to give it up for good.

The peace celebrations of World War I were hardly over when the group of old friends who formed the nucleus of the membership took up the idea (originally mooted by Dr J. B. Lewis and shelved on the outbreak of the war) [3] for the extension of the Show and Sports Grounds to provide for all sports. The proposal was unique and of a nature and scale never before attempted in Australia. The idea gained momentum and no effort was spared in research. Frank Johns recalls that on the first day of his return from active service with the A.I.F. he was buttonholed by Mitchell and Ted Hardie and asked if he had ever visited Gezirek, when in Egypt,[4] and if he thought it within the capacity of the town and district to provide similar facilities.

Representing the Narrogin Picnic Race Club, Mitchell was a member of the deputation which waited on the Narrogin Council on 8 January 1919 to enlist its support in securing sufficient ground adjacent to the present Recreation Ground to provide all sporting clubs, including the Agricultural Society, with a common venue. Other members of the deputation were T. A. Hardie (Agricultural Society), A. E. Hardie, W. F. Bird (Golf Club), G. G. Lavater, E. B. Johnston, M.L.A., and G. F. Parsons. Long and earnest thought had been given to preparation of their case, and one item was worth recording. 'The opinion was ventured by Mr. A. E. Hardie that the ground could also be used as a site for an aerodrome when the use of flying machines came into general use in Australia. For all we know, the idea may not be as nebulous as it appears and such a use may be required of this ground in the very near future.'[5]

The first meeting appeared likely to reach a deadlock, following a councillor's suggestion that the assets of all sporting bodies be pooled. However, delegates accepted the suggestion of the Mayor (Dr D. W. Mackie) that they adjourn and that the sporting bodies again confer with the object of arranging a more definite proposal to lay before the Council.

The Race Club, comprised largely of the same group of enthusiasts who had built it and kept it alive for two decades, was eventually to make the largest contribution from the sale of their course on the Yilliminning Road. The funds thus raised provided for a stand and gardens, the core of the new Sports Ground.

A special meeting of the Council, held on Tuesday evening 13 May 1919, unanimously resolved to 'assist the proposed amalgamation of all sporting bodies with a view to securing a common ground and, that the proposal to raise money by loan or otherwise for the purpose be referred to the rate payers'. No opposition was offered and a provisional committee was appointed. This committee eventually prepared the conditions under which the Municipal Council obtained comparative designs for the laying out of the 'Ground'.

The first Narrogin Sports Ground Board, nominated and elected by the sporting bodies, was appointed in the persons of A. E. Hardie, F. W. Tuohy, C. D. Mitchell, W. F. Bird, with J. J. James as chairman, in January 1920. The first meeting of the Board was held on 14 February 1920, and the secretary, P. M. Sullivan, was appointed on April 14. The Council took over the debts of all sporting bodies in respect of the buildings, borrowed in each instance under guarantee, and opportioned repayment of sinking fund and interest to the original borrower as part rental. At the same time the Council purchased from Michael Brown, at £10 per acre, 50 acres paralleling Clayton Road. This was an extremely generous gesture on Brown's part and reflected his contribution towards the fulfilment of a dream he had long shared with Mitchell: to create within Narrogin a 'Garden City'.

Then commenced a concerted and protracted community effort, equalled possibly only by that of Noah and his family. Kevin O'Connor of Narrogin, now chief inspector for the Metropolitan Market Trust, was of school leaving age at the time. When asked if he had given a hand, he merely replied: 'When you saw Charlie Mitchell and Ted Hardie using a pick or shovel or driving a team and scoop, you just had to hop in and give a hand.' Questioned further as to his impressions of Mitchell's powers of leadership, Kevin replied: 'It's a strange thing that I've never thought of Mitchell as a natural leader, but as one who gave example. To me, he was always the one man I'd met who never made an idle boast nor an empty claim.' In fact, Mitchell had the capacity to attract co-operation in a way that added to the other person's inner sense of importance.

Each sporting group organized its own busy bees and the competition was fierce to rope in helpers. Football ensnared the young parish priest, Father John Lynch, just out from Ireland: not only did he accomplish wonders with a shovel and barrow, but he then decided to learn and play the Australian game, in which, to the consternation of his opponents, he displayed equal vigour and enthusiasm.

Work hardly ceased over the first five years, and as the Greater Sports Ground grew in prestige and community purpose, it made Narrogin famous; it was the admiration of visitors and the envy of every town in Australia. Five years of experience made it patently obvious, however, that the original 90 acres of reserve was insufficient for all sporting clubs' requirements, and in June 1926 the Municipal Council agreed to borrow £2,000 to purchase a further 80 acres from Michael Brown, north of the original acquisition. This was to provide for the Golf Club, which had been established on leased land, and for the re-siting of the football ground.

Only two incidents marred the harmony and spirit of the first two decades. The first, in 1922, was brought about by a division in football control. Frank Barnett, the local dentist, had become the secretary of the Narrogin District Football Association, and advocated playing all matches on Saturdays instead of Sundays, to conform with the decision of a recent referendum that all shops should close on Saturday afternoons to permit organized sport. The Railways and Williams Clubs claimed that there was no agreement to comply with the referendum and, as the majority of their players worked Saturdays, such action victimized their clubs. These two clubs, supported by the Towns Club, formed a breakaway association named Narrogin/Williams Association—with the ardent Frank Byrne of the W.A.G.R. as secretary—and sought registration with the Western Australian Football League. When the Town Council, supported by the Greater Sports Ground Board, voted to bar the playing of football on Sundays on any ground under the Council's control, the recalcitrant clubs promptly arranged to play their matches on Micky Brown's paddock, part of the old polo ground situated north of the then Greater Sports Ground. Each association had its numerous and vociferous supporters, and the controversy as to which association was entitled to registration became known as the 'Battle of the B's'. Byrne sought official recognition of the fact that the greater number of townspeople found their way to Micky Brown's paddock to watch the deeds of prowess the mighty warriors of Williams and Narrogin put forth for their edification. Barnett poopooed the claim, stating the support should be measured by the fact that there were 'more buttons than threepences in the Sunday collection boxes' at Micky Brown's. The problem was finally resolved satisfactorily by the teams joining forces. This was achieved under the wise guidance of T. J. (Tom) Brett, a Western Australian Football League delegate who was shortly to take up residence in the town and who had been invited to act as arbitrator: he had purchased a garage and agency and was also later appointed stipendiary steward for all Great Southern racing by the W.A. Turf Club.

The other incident was a consequence of the depression years, and unfortunately was not peculiar to Narrogin alone. On 19 January 1932 the Municipal Council met a deputation led by W. F. Bird, chairman of the Greater Sports Ground Board, requesting the Council to raise a loan to fund the £1,450 combined liabilities of the Agricultural Society, Bowling Club, Golf Club and Trotting Club, in order to relieve guarantors of their responsibility in respect of overdrafts. Memberships had fallen off, fees had to be reduced, income was static, guarantors who previously had substantial assets now required assistance themselves. It was stated that, to offset this liability, an amount of £3,400 had been spent by the respective bodies in improvements to the grounds, all of which remained the property of the Council. The Council eventually agreed to raise the loan, including in it £450 which the Council had lent to the Football Association, and all the sporting bodies concerned accepted responsibility for repayment of sinking fund and interest on the portions borrowed on their behalf. It is difficult, almost impossible, for the ebullient sporting clubs of this opulent era to appreciate the difficulties which were faced and overcome by similar bodies of thirty-five years ago. Nevertheless, in the spirit still generally associated with the country centres, the finest anodyne to community boredom was community action and fun, and countless stories are told of the simple, human, maybe not very sophisticated shenanigans with which Mitchell and his friends filled the hours which now, perhaps, would be reaped by television or drive-in movies.

When the press announced his brother James's appointment as Lieutenant Governor, Mitchell was away in the country. On his return to Narrogin he called in at the office and at five o'clock made his way to the club. There, to his consternation, he found every available member who could be rounded up (by the irrepressible and then inseparable pair, Lu Cullen and Doug Schoolar) arranged in two lines, hats held to chest, standing stiffly to attention, awaiting his arrival. The 'Court Chamberlain' (Lu Cullen) then led him through the two lines, the members bowing deeply and murmuring: 'Welcome, your Highness!' to the draped presidential chair where he was then enthroned with all due ceremony. Mitchell took the ribbing and merriment in great part. He, the inveterate kidder and leg puller, always conceded that 'if you couldn't take it, you shouldn't give it'.

The club held a barbecue the following year; the menu: beer and chops grilled to one's own taste and by one's own effort. The story goes that after eating fourteen chops, Ron Kennedy of Highbury, a then single farmer of giant stature and gargantuan appetite, expressed himself very feelingly to the organizers: 'This is a hell of a show—a bloke puts in ten bob for a feed and all you give him is hors d'oeuvres.' Mitchell turned to 'Butcher' Brown and whispered: 'We'll fix him. Come on down and open your shop.' Selecting a full rib of beef, they broke the bone in several places, enabling the rib to be coiled in a frypan some 18 inches across, and then proceeded to cook it and present it to the complainant. Ron accepted the rib in the spirit in which it was given, and proceeded to devour it without apparent effort, thanking the organizers for their thought and stating he felt much better. Mitchell turned to Brown again and said: 'The joke's on us.'

'Like hell', said the butcher, 'It's on me. I'm out a rib of beef.'

The club was not all fun and games, however, and like any organization in which Mitchell had a finger, it was where support of worthy civic requirements was discussed and triggered. Even so, the services usually proved an entertainment in themselves, as when kangaroos, increasing in numbers, became a first-class problem to the farmers in the Tarwonga and Culbin districts. Mitchell, keenly aware of the position from observation and conversation, offered to organize Sunday shoots from amongst the club members to assist in eradicating the pests. Ten to twelve guns was the ideal number, strung out in line abreast, with a minimum of five chains apart. The noise of a twelve bore was a veritable whisper to the roar Mitchell gave if any shooter broke the line or spacing. Seemingly casual in all other arrangements, he was adamant on this procedure; he also insisted that every shooter break his gun when getting through a fence, and that on return to camp every gun had not only to be unloaded but had to be broken and remain open. Anyone who offended twice was never asked again.

The shoots soon became publicized and in consequence popular, with more and more clamouring to participate. Mitchell lost interest; while gregarious by nature, he would not take part when clamour overcame caution.

Moreover, further to inhibit his taking part in the shoots, it was at about this time—in 1935—that he was promoted senior manager with his duties centred in Perth. He now spent only occasional weekends at home, and when he did, he preferred to remain within its peace and quiet.

There were special occasions, however. On Thursday 15 October 1936 the Lieutenant Governor made his first official visit to Narrogin to open the Show. He was a keen and excellent shot who never lost his love for hunting, and as his new and exalted position made it increasingly difficult for him to indulge his liking, a special shoot was arranged in his honour to take place in the Culbin area on the following Saturday. Charles naturally was a prime mover in arranging the outing.

J. A. ('Tiger') Scott, now living in Como, was among the party of twenty who left Narrogin just before noon. He recalls that Sir James, because of his age and a temporary indisposition, was given the seat of honour on an ant-hill while the balance of the party, divided into two halves, travelled by truck to the north and south and in two long wings drove the game back toward him. By 3.30 p.m. the main body had accounted for forty 'roos and Sir James, to his great satisfaction, had managed another five.

There were then 'refreshments' to be disposed of, followed in due course by chops barbecued on a piece of rabbit netting, plus the inevitable 'spuds' baked in the ashes. To celebrate his elder brother's visit Charles Mitchell had killed and dressed a whole sheep, which was hung from a near-by wandoo tree. Each member of the party was expected to carve his own particular joint with a sharp axe and a butcher's knife stuck, for convenience sake, into the bark of the tree.

Joe Monaghan, the proprietor of the Hordern Hotel, had requested the honour of supplying a hamper for the Lieutenant Governor, whom he considered could hardly be expected to eat messy barbecued chops with his fingers and hold hot baked spuds in his hands: custom at the shoots made the use of knives, forks or plates strictly taboo. Predictably, Sir James ignored the chicken in aspic and the salads, and headed for the chops and spuds. By late afternoon he had as much grease on his fingers and face as the next man.

It was at about this time that the uninhibited Lu Cullen challenged Charles Mitchell: 'You've always boasted you're a great shot on the wing, Charlie', he said. 'Let's see just how good you are.'

'Right', said Mitchell, loading and cocking both barrels. 'What do you want me to shoot at?'

Lu produced Sir James's celebrated bowler hat from behind his back and tossed it high into the air. One shot blew the centre out—to the great delight of all present, even Sir James, whose grimy face wreathed in smiles.

'Charles', he chuckled, 'you haven't improved.'

CHARLES MITCHELL CONTINUED AS A MEMBER of the Greater Sports Ground Board until the committee meeting, 13 November 1935, when he advised that, owing to his pending transfer to Northam, he would not be eligible for re-election at the Annual Meeting. It is recorded that, following his retirement, he attended the subsequent Annual Meeting of 30 November 1943, seconding the receipt and adoption of the chairman's report, and later recommending to the Board that at the earliest opportunity couch grass be planted on the race track from the entrance into the straight to the seven-furlong starting post.

Many years later he was invited officially to open the memorial gates leading into the Sports Ground, which were donated by Aubrey Hardie in memory of his late father and uncle, Tom and Edward. A brass plaque on the gates reads:

> These gates were dedicated by Aubrey A. Hardie to the memory of his Father, T. A. Hardie, and Uncle, A. E. Hardie, who enjoyed many years of sport on this ground. March 1953. Unveiled by C. D. Mitchell on 26th September, 1953.

As the Sports Ground was officially opened in October 1921 by the then Premier, James Mitchell, it was most appropriate that his brother Charles, who had played such an active part in its creation, should now perform an opening in memory of those contemporaries whom he had survived.

Mitchell accepted the invitation with alacrity—gone were his shyness and reticence. His speech was the highlight of the day; for once, a rare occasion, he spoke feelingly of people he loved, and of the project, so dear to his heart, to which he had contributed so much.

Mitchell served on the Committee of the Narrogin Club for many years and as president for the year 1925-26. A recommendation for Life Membership made at the Annual Meeting, 26 July 1940, was confirmed at the General Committee meeting, 21 August. In the Narrogin Club his photo is hung along with those of fellow presidents, but he will always live far more vividly in a still widely used expression he coined for a poor shot at billiards and snooker: 'a gin of a shot'.

He will be remembered at the Race Club by a memorial plaque to be placed on the gateway leading from the enclosure to the saddling paddock, inscribed:

In memory of C. D. Mitchell
'Mitch'—a true sportsman
who worked so hard to help establish
this Greater Sports Ground

1. A residence leased from John Barron. Barron called a meeting of creditors on 6 July 1915, and on the following day assigned all his estate to Solicitor W. F. Andrews, who sold the property to the club for £700. Transfer was registered on 19 August 1915. It is now used as a doctor's surgery.
2. Australian whisky was first marketed in the late twenties under the trade names of Brinds, Tolleys and V.O.S.
3. On 11 May 1898 an area of 10 acres, Reserve 1836, in the south-east corner of the present Reserve 14387, was declared and vested in J. T. Chipper, M. Brown and A. Trefort for the purpose of 'recreation'. In 1904, to permit the Agricultural Society to transfer their activities from the corner of Egerton and Earl Streets, Reserve 1836 was cancelled and included in an area of 23a 3r 10p, Reserve 9456, vested in the Recreation Ground Board, comprising William Graham (Snr), F. C. McKinnell, W. F. Wiesse, J. E. Clayton, T. P. O'Connor, W. L. Graham and R. V. Uren, for the purpose of 'recreation and Show Grounds'. Led by Dr Lewis, president of the Agricultural Society, representations in 1912 led to the Municipal Council agreeing to assume control of the ground, the Board of Control having given its consent to the change. Reserve 14387 was set apart and notice appeared in the *Government Gazette,* 11 October 1912. Reserve 9456 was cancelled.
4. Gezirek Sporting Club was situated on a small island in the River Nile. There was no form of sport played by English people which was not provided for. The social side was catered for with cafes and restaurants located alongside swimming pools. During the Second World War, the Australian Army had two houseboats tied stern to stern at Gezirek, one used for personnel on leave, the other for convalescing wounded.
5. *Narrogin Observer* January 1919.

18

Exodus and Retirement

This time Charles Mitchell took care that news of his transfer to Geraldton was known only to the immediate family. There were to be no more send-offs or farewells. So instructions were left with land agent C. W. (Cyril) Allen to store the furniture and let the house. Like the Arabs, Mitchell and his wife 'folded their tents and stole away'—on the afternoon of Sunday 2 January—to reach Geraldton in easy stages for the opening of business on the following Tuesday morning.

The *Narrogin Observer* recorded in the personal column, 8 January 1938: 'Mr. & Mrs. C. D. Mitchell left for Geraldton, to which centre Mr. Mitchell has been transferred. They have long been resident in Narrogin and will be missed by a wide circle of friends.' What a different exit from that of September 1935!

As no official quarters were provided for branch managers, the Mitchells lived at Jack Shea's Club Hotel from January 1938 till early December of the same year. They then were able to move into the home of Harry Heaton, the branch accountant at Geraldton, who had left on three months long service leave.

However, much as he tried to hide his feelings, Mitchell was disappointed and hurt: more in that he had to leave Narrogin than in missing the compensating appointment of chief inspector (wheatbelt). In his fifty-ninth year he had to make a new home far from his beloved Narrogin.

Even so, there was one bright spot. His son-in-law Colin Wishart was on the staff of Elder Smith and Co. in Geraldton, so that he and Ida had the delight of the company of their daughter Joan and their grandchildren, Michael and Janet (Topsy). On the other hand, the family had broken up their home for the first time, a disturbance which brought Peggy to Geraldton to work and to live with her mother and father at the Club Hotel.

One of the first things she did, and possibly it was to relieve Mitchell's sense of frustration or of failure, was to purchase from a farmer in the Greenough district a grey horse regarded as a rogue. Jack Shea was only too happy to stable it. Because of the grey's reputation, Mitchell considered that he alone could manage it—which was exactly what had been intended. A horse could always be borrowed for Peggy from Mrs McAleer's stable, where at any time there were five or six horses needing exercise, and together Mitchell and his daughter entered upon a series of memorable weekends, riding for miles and exploring every inch of the environs of Geraldton. Peggy's thoughtfulness was well rewarded. Her love of horses brought her in contact with her future husband, Walter McKenzie Grant, a distaff grandson of Maitland Brown, with whom Peggy's paternal grandfather had had business associations in the 1870s.[1]

FARMERS IN THE EASTERN PORTION of Mitchell's new territory had gained a reputation for militant individualism during the depression years, and Mitchell's friends lost no time in warning him, in a 'friendly way', of how they treated bank managers out east. He'd better look out, especially in Morawa!

In 1927 Morawa was described by the Western Australian newpaper *Truth* as 'a small country town surrounded by scrub and beer bottles'. The following year it achieved the dignity of having its own hotel, as well as one trading bank which it very nearly lost shortly afterward.

Some time before Mitchell's arrival, a farmer had arrived in town to importune his bank manager for an advance of £10 for stores and parts. Despite his earnest pleas, the request was firmly declined. Fed up, broke, and completely disheartened, he set off for the nearby pub. There he poured out his tale of woe to sympathetic listeners. They could do no more than tell him how b—— sorry they were, and press him to have a drink. The consolation lasted through the entire afternoon until, by dusk, the aggrieved farmer had come to a decision. Taking a number of empty beer cases and a heap of straw packings from behind the hotel, he piled them up against the side of the Bank, but just as he was about to strike a match he was collared by the local policeman who, in all probability, had been tipped off.

When he was arraigned next morning in court (the local hall) before two of the local justices, the male population of the town was arrayed on one side of the chamber against the bank manager, in solitary state, on the other. After the constable's evidence and the plea: 'I suppose I'm guilty!' were heard, the justices conferred. No matter how sympathetic they might have been personally, they were also debtor customers of the Bank. Their decision: £10 or fourteen days. A noise, not unlike a growl, rose from the public section. The prisoner looked at his bank manager, who stared fixedly to his front.

'What about it?'

The manager shook his head.

'Alright, if you won't pay my fine, you'd better get somebody to caretake the place while I'm in clink.'

The banker was in a quandary. To arrange surveillance of the 'prisoner's' farm would cost far more than payment of the fine, and, moreover, some sort of public discomfiture was certain to ensue to him from whatever the prisoner's friends could be relied upon to cook up. He thought quickly.

'I'll pay the fine', he announced, and the crowd murmured its approval.

'Well, now that you've opened your heart, what about the £10 I want for stores and parts?' the prisoner demanded, doggedly. 'Or do we start all over again?'

Challenged publicly the bank manager was left with no option: 'Alright', he said, and his capitulation was greeted by light cheers and hand clapping from the public. Justice was served, peace was restored, and all adjourned to the pub, where the bank manager was made welcome and allowed to buy the first round.

Mitchell, if anything, looked forward to meeting these characters. He enjoyed 'hard cases', provided they had a sense of humour and an appreciation of 'fair go' as exercised toward them and by them.

Another story of Morawa, not necessarily about the same bank or manager, is told by James Bligh, of whom we have read earlier in the chapter on the Depression.

Bligh had tendered, without deposit or offer of collateral, for an abandoned property situated between Morawa and Three Springs for the bank debt outstanding. His offer had been accepted, the bank being only too glad to be able to protect its investment, but the local manager insisted on Bligh's completing a stock and station mortgage, which would give the mortgagee a charge over all free items of stock and plant, as opposed to the simple overdraft mortgage against the land only, which Bligh had proposed and which he had firmly believed was all the Head Office would want.

The argument had proceeded through December to the end of February, without either party giving way. Bligh could wait no longer, as, in readiness for the coming season, he wished to commence cultivation on the new property, which comprised heavy land in a proven rainfall area.

It was now time to produce his trump card. Boots and socks were not only expensive items of wear, but they were also not as comfortable as a pair of sandshoes worn without socks. Bligh had worn his sandshoes throughout the whole of harvest and wheat carting and, under such circumstances, apparently, the correct proportions of perspiration, heat and dust act as a catalytic with the rubber to produce an odour which has to be experienced to be believed.

Reviving the olid aroma by neglecting to bathe his feet for a couple of days beforehand, and making sure that he had perspired fairly freely on the day itself, Bligh presented himself in the manager's office before closing time to settle finally the question of which type of mortgage he would sign. He seated himself without being asked. The manager visibly blanched, his nostrils quivered, but years of rigid training and attention to the Bank's interests stood him in good stead. He was not to be deterred. Bligh pulled his chair nearer the table, and pushed his feet forward as far as he could. This was just too much for the manager, whose gorge rose.

'Get out of my office, Bligh', he said.

'Not until I've signed the ordinary overdraft mortgage', was the reply.

They glared at each other, and perspiration poured from the manager's face. He stood another three minutes of the atmosphere. Bligh was not wholly inured, but the shoes were by now closer to the manager's nostrils than to his. There had to be a breaking point. It came, but in the defences of the Bank.

'Alright!' the manager agreed. 'Here's the mortgage you want. Sign it and get out of my office and don't come back for at least six months. And burn those sandshoes!'

TWO IMMEDIATE CHALLENGES AWAITED MITCHELL'S attention in Geraldton. The first was to make an appreciation of the soils and agricultural potential of the Northampton-Ajana-Yuna areas, which he later described as 'good enough to eat', and which later influenced him to give those settlements his utmost support.

'Us Fellows' Football Club

Players, supporters and 'lady-servers', with 'grandstand' at rear

Charles Mitchell with grandchildren

Philip Rossiter Mitchell
(Ginger), 1941

Charles Rossiter Mitchell
(Bricky) and
Ida Marian Mitchell, 1941

The second was to define by boundaries the 'safe' and the 'non-safe' wheat-growing areas in the northern region of his district. A prerequisite to Commonwealth government assistance, this task was undertaken also on behalf of state financial institutions which were pressing for an official definition. Their 'pockets had been hurt'; they required protection in future lending and that a warning be issued to future settlers.

Accompanied at every available opportunity by his accountant, Harry Heaton, Mitchell travelled extensively to study conditions at first hand in regard to the second challenge. The evidence of isoheyts and cropping statistics became a minor consideration. Visual observation was a sufficient basis for Mitchell's determination of the ruling geophysical boundaries, and on that basis a line was drawn from the coast, at the 28th parallel, due east through Ogilvie, then in a sweep to pass through Mullewa and Gutha, continuing in a south-easterly direction to the limit of Mitchell's responsibility at Kalannie.

Employing Mitchell's commonsense formula as a basis, W. Wardle, as the Bank's chief valuer, subsequently extended the line to the Southern Ocean at Culham Inlet, west of Hopetoun (see Map 3), a task in which he was assisted by his close friend, S. E. Smith, recently retired Lands Department district surveyor (Northam). With minor adjustments this practical determination has stood the test of time, except that, predictably, the farmers east of the 'Marginal' line soon took objection to the term 'Marginal areas'. They were somewhat mollified by the application of the term 'outer stock areas' in lieu.

Prior to 1938 reconstruction of the outer wheatbelt was no more than a piecemeal operation, dealing mainly with the 'special areas' of the Lakes districts, Southern Cross, Bullfinch, Ravensthorpe and Salmon Gums. Emphasis was on linking farms, on the writing down of Agricultural Bank advances and adjustment of land rents, losses eventually being admitted by the Commonwealth Grants Commission. There was, however, little fresh capital available for extension of and consolidation of improvements.

Reconstruction in its full sense became effective late in 1938 when federal and state legislation, establishing a consumption price for wheat in Australia, permitted the Commonwealth government to provide special financial assistance to the states for wheat grown on Marginal lands. Finance became available under the Wheat Industry Assistance Act, No. 53 of 1938, 'for the reconstruction of marginal areas within an approved and clearly defined line, conditionally upon the State's agreeing to the area that was to be sown or fallowed in any one year and ultimately the prohibition of wheat growing for grain within such areas'.

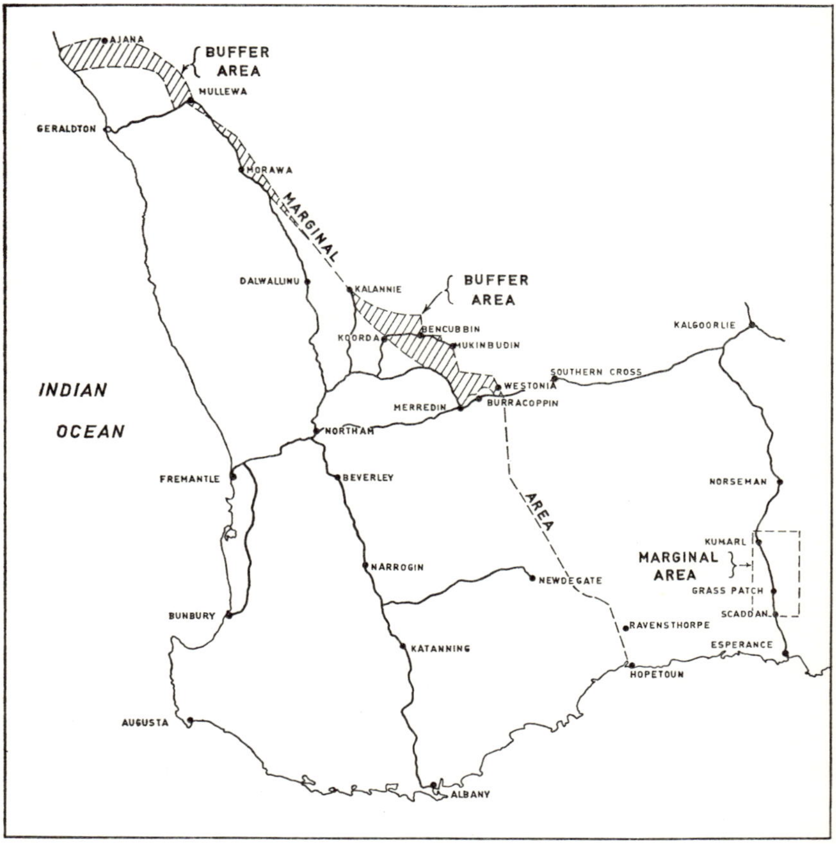

Map 3

Under the provision finance was made available to the four states[2] in which somewhat similar problems had been created, primarily, by alienation of Crown land for wheat growing, before related economics and farming techniques had been proven.

Thanks to the foresight of the Board of Commissioners of the reconstituted Agricultural Bank and to the lead given by Mitchell's initial field reports, an estimated 75 per cent of Western Australian Marginal areas (as represented mostly by the 'Special

Areas') had been examined by the end of 1938. The state was thus able to enter into immediate agreement with the Commonwealth that its plan for reconstruction would proceed:

(*a*) by linking of properties and further development, and
(*b*) by transfer of selected farmers to better rainfall areas.

In respect of (*a*) funds were found for the following purposes by way of grants:

1. Sheep numbers or discharge of existing bill of sale to stock firm, maximum £300.
2. Additional clearing where necessary to provide a full strength unit (originally regarded as 1,500 acres of cleared land but later increased to 2,000 acres).
3. Fencing, sufficient to ringfence and subdivide the cleared areas.
4. Provision of dams or bores.
5. Pasture seed—Wimmera Rye.
6. Sundry fees, including Lands Department application and survey fees.

The idea underlying the reconstruction was that a change should be made from wheat growing solely to sheep raising, or a combination of the two. The Agricultural Bank agreed to reassess the writing down of capital debts on all Marginal zone properties to an amount which, it was considered, those various properties were able to carry.

The Lands and Surveys Department, in making their contribution to the realistic appraisal of the problem, defined all alienated leasehold land in the Marginal areas as 'Grazing Lands', thus enabling farmers to hold up to 5,000 acres of leasehold country, irrespective of quality. The greater part of the land was then repriced at one shilling per acre, the minimum permitted under the Land Act.

The Lands Department's boundary for the minimum price zone was not exactly the same as that of the Marginal line drawn by the Agricultural Bank. For good reason, the Lands Department line followed existing survey lines wherever available and applicable.

In consideration of the assistance described, the farmer agreed to restrict wheat growing to 200 acres a year. The drought years of 1938-39 and 1939-40 prompted the Wheat Stabilisation Board to suggest that wheat growing (by this time carried on under licence) be limited in Marginal areas—and subsequently permissible wheat planting was reduced to 150 acres.

The growing of wheat, within the defined Marginal areas, although later lifted to a maximum of 250 acres, continued under restriction. The year 1949 saw a rationalization of individual limits, proportionate to total area cleared but within the five-year rotation, which was designed to encourage and preserve stock feed.[3]

This formula did not apply to all wheat farmers. The Marginal Area Reconstruction Board ruled that the State Wheat Stabilisation Committee should not limit wheat growing to Associated Bank clients—the limitation applied only to those Agricultural Bank clients requiring seasonal and/or developmental assistance.

By 1954, influenced by increasing world wheat markets, restrictions were forgotten and were only a matter of record until revived in 1969-70.

Although by 1935 the number of sheep held by Agricultural Bank clients was increasing steadily, the condition of the flocks revealed that most farmers had little experience, and were in urgent need of advice on husbandry and management. Consideration had to be given, also, to the pressing stocking needs of the outer wheatbelt districts (the Lakes areas, Salmon Gums, Southern Cross and Bullfinch), which were subject to reconstruction programmes and in which the Commissioners had adopted Mitchell's recommendation that farming could be profitable only where sheep and wheat were grown together.

The Commissioners rightly required that money for purchase of sheep should be advanced only on the advice of some thoroughly reliable person; but where was such a person to be found? The bleak outlook was relieved by a fortuitous coincidence in which the services of Charles Pendergast Murray[4] became available. Appointed to the Department of Agriculture on 12 March 1936, an adviser in sheep breeding and management, Murray was promptly seconded to the Agricultural Bank. His appointment was timely, but there were many prejudices to overcome, particularly among stud breeders, to whom he became the bane of their lives. Each considered his own flock not only perfect, but sacred. Stock agents and advisers felt that there was little they could be taught. All resented criticism and unsought advice. Murray had just as fixed an opinion about sheep types as he had about people; he was never one to mix or use soft words, and unfailingly made his point publicly, to his opponents' discomfiture.

Mitchell was an excellent sheep man himself. He recognized Murray's tremendous knowledge and capacity, and was one of the few who accepted him readily and wholeheartedly. For some unknown reason, however, there was a clash—possibly one of personality, since each was a rugged individualist. From Murray's side the breach was apparently irreconcilable.

Nevertheless, although Murray studiously avoided Mitchell's company, Mitchell never for one minute withdrew his enthusiastic support and advocacy of Murray's talents, nor lost his own delightful sense of humour.

Tom Jones recalls being told by Mitchell, when acting as accountant at Kununoppin branch during the months of Mitchell's term as relieving manager: 'There's a truck load of rams Murray ordered just come off the train. Ring him and tell him about them, and mention that it will be alright to come over tomorrow as I'll be away in the bush!' Tom looked at his boss quizzically and was informed: 'The silly old b—— won't come if he knows I'm here!'

Murray was tall and extremely spare of frame, and his appearance is best described in a story he used to tell against himself. 'I tried to join up in the First War. The recruiting sergeant studied me for a while and then told me to clear off, as I wouldn't be of any use other than as a pull-through for a big gun.'

I went out with him on many occasions on flock culling and selection. He was a good bushman and lived hard, harder than anyone else I can recall. He always insisted on hosting from his own tucker-box and, whenever I was with him, either by design or coincidence, the bread was always stale and more easy to break than cut; butter, if any, was rancid; sometimes he brought along a tin of sardines, but always a great hunk of cheese, which had been pared until only about half an inch adhered to the rind. I never saw him produce a whole piece. Strong black billy tea topped off this spartan meal and more often than not Murray forgot to bring any sugar for the visitor, for which he apologized, often with a very deep twinkle in his eye. Like Mitchell, he loved a few whiskies and conversation with intimates, and found relaxation of an evening over the camp fire.

This man had an instinctive genius in dealing with sheep. Even then, so long ago, he spoke of the advantages of late summer and autumn shearings, and suggested that breeding sheep with woolly faces were inherently neither as prolific nor as 'good doers' as their counterparts with plain faces. It was some thirty years later before these theories were proven experimentally.

I doubt if he was ever accepted by stud breeders as a class; perhaps his wisdom and his principles were adopted, partly or wholly, but without acknowledgement. I recall his telling me after some six months close association with him: 'Gabbedy, you'll never be a sheep man as long as you live, you just haven't got it.' My face fell, but then came the solace: 'Don't let it worry you. Even as it is, you know more than half the stud breeders!' I didn't believe him then and I don't now, perhaps less, after thirty years of experience: but I survived on the thought that I could get on with Murray.

He clashed with officialdom and, perhaps to save face, they gave him the minimum of direction.

There was an indigenous weed, which grew rather prolifically in the Ravensthorpe and Salmon Gums districts, known as shepherd's lucerne (*Lepidium hyssopifolia*). It had been introduced, possibly, from the eastern states during the gold rush days, when Esperance was used as a port of entrance. The Department of Agriculture did not look with favour on shepherd's lucerne; it was not particularly palatable and difficult to control, except by cultivation. Murray, however, liked it and in his travelling always seemed to have a tobacco tin of seed in his pocket, which he surreptitiously distributed, explaining to the farmer that 'Yarrum' grass would provide excellent feed. Few seemed to realize that 'Yarrum' was his own name, spelled backwards.

The story is told of two young Agricultural scientists, close friends of Murray's, coming across him at the Royal Show leaning over the pen of the champion merino ewe.

'A beautiful animal, Mr Murray', said one.

Murray took a while to reply. 'It should never have been given first prize.'

'Why?'

'Get into the pen and find out for yourselves', said Murray.

They recall going over the ewe as with a fine tooth comb; it was perfect in every respect until, as the last procedure, they inspected its mouth. The animal had a black spotted tongue. Somewhere in this stud-breeder's line of breeding he had introduced an English ram to strengthen or eliminate some defect in a selected merino flock which was then bred back to pure-bred merinos. The final standard of perfection, whilst acceptable to the Show judges, could not be tolerated by Murray. A black tongue was to him something a pure-bred merino would never have. A further point to the story was that Murray's two friends were certain that Murray himself had never mouthed the sheep; somehow or other he knew just from observation.

Murray was employed by the Bank in his original capacity until his age retirement on 4 February 1950. He was then re-engaged in a temporary capacity till 31 July 1951. During his service with the Bank he personally selected and purchased 150,000 ewes in the wool (at an average price of seventeen shillings) and approximately 3,000 rams, and arranged for delivery of all to 780 farms stretching from Northampton to Salmon Gums. When his purchases totalled 90,000 sheep, it was recorded that deliveries were only ninety short—a truly remarkable achievement of which Commissioner C. L. Clarke said at the time: 'I feel sure its like has never been accomplished anywhere else in the British Empire.'

Drought conditions experienced in 1938-39 and 1939-40 greatly hampered the reconstruction scheme. While advances had been made for fencing and for dams, the purchase of sheep was restricted by the absence of feed and because the rainfall was not sufficient to provide run-off water to fill the dams. However, the necessary sheep purchases had been virtually completed by 1943, and the next eight years were devoted to supervision by the Bank's officers of flock management through culling and ram selection.

By the time he was due to retire Murray had the satisfaction of having played a big part in transforming the district in which he had operated. A vast area of country opened up specifically for wheat growing had failed at a cost of millions of pounds to the government. Over the course of fifteen years this same area of country had been converted into pastures carrying a million sheep, and the settlers who had remained were secure and prosperous. Western Australia was truly fortunate in securing another such 'man for the times'. It would not be possible to applaud Murray's work too highly, or even to measure the benefits which accrued from his wise judgements; from his selection of ewes and rams for foundation flocks and of typed sheep for various districts, and from his advice generally on sheep management and wool production.

THE DEMARCATION LINE FOR THE MARGINAL AREAS had hardly been defined when it was realized that an anomaly had been created with respect to the areas either side of it. While land west of the line, that is, outside the Marginal area, theoretically retained its value, that east of it dropped to as little as one-fifth of its former worth, a circumstance all the more aggravating to the owner of the devalued land in that the line of demarcation was often not a physical or climatic consideration but merely a man-made road (see Map 3).

As a result of solid ground work by Chief Valuer Wardle (later to become assistant manager), and through the support and persistent representations to the Commonwealth government of R. W. Brownlie (secretary, and later commissioner), the Commissioners of the Agricultural Bank were able to define and to have accepted 'buffer areas' of varying depth on the western side of the Marginal line to which the benefits of reconstruction were also available in varying degrees.

While the big wheels concerned with government policy had been whirling publicly on behalf of the wheat and wool industries, the smaller cogs had continued to mesh, generally unnoticed, in their narrower spheres of influence, otherwise there would have been little industry for the government to deliberate about.

The most important of the smaller cogs were the field inspectors who made up most of Mitchell's world. The inspector had to be available at any time to deal with the 'higher-ups', to instruct and inform them of matters at a level to which time and their commitments did not permit them to penetrate personally, but he had also to face up to the unenviable responsibilities of the man-on-the-spot during the Depression, in most cases as controller of not less than 75 per cent of the properties in his district, banker, caretaker, confidant, lawyer, accountant and, often, final arbitrator in matters which might be well outside the scope of most laymen.

The 'Bank-man's' home was invariably his office. Clients would visit him there at all hours, after the pub shut, on Saturdays and Sundays. Inspectors from various areas rarely met to swap experiences and to compare notes: there was no contact with Head Office, and communication with District Office staff was on an impersonal basis.

A classic example of isolation and adaptation was that of Inspector Lew Whitford, who was stationed at Harrismith, 42 miles east of Narrogin. The town consisted of a school, a store, and three houses. The twice-weekly train from Narrogin, with mail and fresh meat and vegetables, arrived at 11 p.m. Tuesdays and Fridays. Everybody waited up for it. Motor cars were few and far between, and without exception, some Agricultural Bank clients who had either been to Narrogin office or to Head Office, got off the train and wanted to see the local inspector. Lew's office, on the side verandah, closed when the last client had departed for home—anything up to 2 a.m. While they waited patiently for their turn, Joy, his wife, served them with cups of coffee.

At Newdegate, Friday was shopping day, and it was recognized that the inspector—during harvest and seeding seasons—should always be in town on that day. One was very fortunate to finish before midnight. I know, because I spent three years there as inspector. The near-by hotel served as a waiting room, with the client who had finished his business going back to the bar to tell the next in line it was his turn. There was no need to serve coffee—clients at this hour of the night were always amply refreshed. The coffee drinkers had left for home before dark.

Naturally enough, the field and office staff attached to the Geraldton branch welcomed the advent of their new boss, whose reputation for hard work, fairness, geniality and accessibility had preceded him up the line.

Harry Blake, an inspector then stationed at Perenjori, isolated, lonely, and at times uncertain how to proceed, recalls Mitchell's periodic visits as 'a breath of fresh air that lasted for weeks afterwards. He'd listen to your worries, think about it, and deliver one of two favourite pieces of advice', Harry recalls. 'Either: "You'd better give him another go", or: "Have nothing to do with him—he's a bloody head." ' To be so described by Mitchell put a man in the category of 'con-men', almost—definitely among the 'lurk merchants'. It wasn't always as simple as that, however.

Harry Blake recalls one famous occasion on which Mitchell was routed by the reputedly fair sex: the distaff side of a 'no-hoper' couple farming east of Latham, whom Mitchell knew by reputation but had never met. They had long resented Blake's efforts to dispossess them, and Mitchell and Blake drove out for another try. The wife sighted them coming and when they were still at least a hundred yards from the house she began to abuse Mitchell in most unladylike terms. Mitchell the bull, the fighter, the rough-rider, was very shy in the presence of women. He never voluntarily sought their company and, as on this occasion, generally accorded them a degree of gentleness many of them neither deserved nor practised themselves. His expostulations went unheeded, and she was still showering the two Bank men with abuse and shocking language when they retreated to their car twenty minutes later.

'Mitch was visibly upset', Blake recalls. 'When we got under way he said: "That's a terrible woman, Blake. Have nothing more to do with her. Just let them be." And that was his final instruction.'

Because of his earlier association and investigations Mitchell returned to the Moorine Rock district during the winter of 1939 to assist and advise George Gillies, the Northam branch manager, in the reconstruction of that area. As George recalls to this day, Mitchell did the planning and he wrote the reports, which Mitchell, nevertheless, insisted on checking; they had to be just right. 'I was most impressed with his knowledge and practical approach to administration', Gillies says, 'but his routine in field work took a bit of getting used to. When you got used to it, it was wonderful, but imagine how you would feel when the boss would insist on camping out—even in the rain!—when there might be an empty school-house only a few miles off.'

Mitchell loved life. There were few aspects of it from which he did not extract pleasure, but there were some which were never failing sources of delight—children, poker, horses, solo whist and his favourite brand of Australian whisky. He owned a Ford utility, and his favourite occupation at weekends and on holidays was to pick up a load of kids, take them down to the beach, buy them icecreams and keep a fatherly eye on their games and paddling. He had real understanding of children, and never tired of telling them stories full of native humour and inventiveness. The recollections of some of the children, now parents still living in Geraldton, recreate the scene as he sat on the beach there, children clinging to his hands and his words while he related again and again their favourite story, 'The Pig with the Twisty Tail':

> Once a long time ago in Bunbury I had a pet black pig, with a lovely shiny coat and the longest, twistiest tail you ever did see. This pig was taught to pull a little cart and take children for a ride; but he was a naughty pig and if Mitch took his eye off him for only one minute, he would head for the nearest pond and run into it. Then he would heave his right shoulder and tip the cart over. The children would all splash into the water; the boys would love it, of course, but the girls would all squeal with fright. Then the little black pig would right the cart and wait on the bank to take them back to Mitch.

The story was told so well and so often that Peggy, who left for boarding school at the age of thirteen, boasted there of her 'father's pig'—to the great delight and guffaws of her classmates. Bricky, in true younger brother style, enjoyed teasing Peggy about her credulity, and on the birth of her first son, he telegraphed: 'Congratulations. Hope he cuts his teeth on twisty tail.'

The group of five solo whist addicts who played weekly at each others' homes comprised Mitchell, Bill Brede (stevedore), Bruce Wilkinson (manager of Dalgety's), 'Chappy' Chapman (chemist), and George Brown (clothing store). One of their favourite pastimes was kidding Mitchell about his depraved taste in whiskies. He retaliated, when he was host, by serving nothing but his beloved Australian whisky—but in a variety of Scotch bottles. To the day of his death he would recall with the greatest glee the evenings he spent listening to their learned dissertations on the Scotch he had provided, and their tributes to their host's generosity.

On Harry Heaton's return from long service leave, the Mitchells returned to Shea's Hotel and stayed there till the following June, when their son-in-law Colin Wishart became the manager of Elder Smith's Geraldton office. Colin and Joan moved into the staff house in Marine Terrace, and the Mitchells took over the home Colin had rented in Shenton Street.

The winter of 1940 was a particularly dry one and by the spring of the year near-dought conditions existed in that section of the north-eastern wheatbelt which fell under Mitchell's administration. A plague of grasshoppers accentuated the shortage of natural feed for, while control measures had been introduced, they were as yet still far from effective.

The decade of the thirties brought rather more than a fair share of Mother Nature's caprices to the north-eastern section of the wheatbelt, and I believe that this history would not be complete without reference to one of the most curious of them, which raised a tremendous pro-and-con hullabaloo at the time, but which has since been completely forgotten; I refer to the seldom-told story of the great 'emu war'.

The offensive launched against the emus on 2 November 1932 was probably unique, as a military operation, anywhere in the world. A shortage of feed and water brought about by dry conditions in the Murchison forced an estimated 20,000 of the big birds to migrate to the south. They followed the No. 1 Rabbit

Proof Fence to where the northern edge of farming civilization ran east and west through Campion.

The big drawback in the practical application of the Fence was that so many people neglected to close the gates after passing through it. The emus lost no time in following the rabbits through these providential breaches in the wall, some going west, the remainder spreading east to the Walgoolan district. Although they feasted mightily on the ripe wheat, the damage caused by the banquet was negligible to that wrought by their huge splay feet when disturbed and chased from the table.

Eventually local representation was made through the state government to the Defence Department seeking the assistance of machine-gunners in exterminating the pest. The Department of the Army intervened eventually on behalf of the farmers, but heaven intervened more swiftly for the emu. Before the armed forces could gather at the starting line, heavy rains, filling dams and water holes, scattered the big mobs far and wide in small companies of guerillas, now well watered and very considerably recovered from their previously weakened condition, and living very comfortably off the country.

When the armed party reached Campion, after having travelled by train to Burracoppin and from there by motor transport, every settler within 30 miles of the siding was on hand to meet the punitive party from the 7th Heavy Battery, Royal Australian Artillery, Major G. W. P. Meredith commanding, assisted by Sergeant S. McMurray and Gunner O'Halloran. In a curious fashion forecasting one aspect of the conduct of the war which was waiting at the end of the decade, the operation was covered by newspaper reporters, photographers and a Fox Movietone cameraman and crew.

While the settlers were conferring at Campion siding on the plan of attack, a mob of about forty emus was seen on a crop not half a mile away on a slight rise. Major Meredith decided to give an immediate demonstration of the effectiveness of the Lewis gun fire. Beaters were sent out, but the birds became alarmed and began moving off at a great pace at a tangent to the gun position. Sergeant McMurray opened fire at 1,000 yards and a great cloud of dust appeared just short of the objective. The next burst of fire from the gun struck the brown mass of scurrying birds almost as they reached the protection of some timber and about a dozen toppled over and over as they fell. It was fine shooting at such a long range with Mark VI ammunition and the settlers were very impressed with the display of marksmanship. However, this was to be the first and only real success.

M. E. (Max) Brinkworth, Agricultural Bank inspector stationed at Campion, had been deputed to act as guide, and the party eventually got under way with a procession half a mile long. There was little fencing to direct the movement of the birds, and as they scattered, it was now impossible to carry out the intended drive by several hundred beaters over a 20-mile front to bring them to the mouths of the guns.

Long-range shooting at small mobs was soon recognized as a high wastage of the 10,000 rounds of ammunition provided. It soon became apparent that unless hit in the hip-bone or the heart, the emus could carry quite a number of shots without any undue loss of range or mobility. Its capacity for lead carrying was verified by examination of a bird run down by a truck on 18 November: five bullets were found, some of which had evidently been in its body since the first day's operation.

The spread of fire from the Lewis guns at any distance over several hundred yards yielded such negligible result that the exercise became a laughing matter with the farmers, who had hoped for miracles. They dispersed to their homes, losing interest and later denying an implied community responsibility. One farmer, who had been prominent in representations that the machine-gunners be brought into the district, was later asked to comply with the provisions under which they were made available, and to pay £24 as cost of ammunition expended. He responded with a counterclaim for £924, in which the major item was £900 for '6,000 bushels of wheat at 3s a bushel, eaten by enemy during official delay in opening counter-offensive'.[5]

In relating his experiences during the campaign, Max Brinkworth said that the tactics of the attacking forces provided for ambush positions behind convenient dams so that when the thirsty birds came in at evening they would, in theory, be met with a withering hail of fire and eventually be reduced to choosing either thirst or lead-poisoning as their passport to paradise.

On the second evening, Max's Browning .22 automatic rifle brought down six of the guerillas—one more than the Lewis gun—before they retired to prepared positions. However, in the event of there ever being any necessity to mount a similar offensive in the future, two valuable lessons were learned in the fighting around Campion: one was that the Lewis would almost invariably jam just when it looked like having a chance of wreaking most damage on the enemy; and the other was that its burst exerted the best effect astern, when the birds would turn tail and run, almost invariably in a compact group.

Despite discouraging results, the campaign continued until early December, when it was disbanded to permit the troops Christmas leave at home. Nevertheless, Major Meredith and his party literally ran themselves ragged. Every subterfuge was employed, every trick exploited, traps were built and drives organized. In the end, although no official figures are available from either side, military pundits suggest that it is doubtful whether there were more than 300 enemy casualties. On our side, of course, there were none at all.

The war excited the interest of the local press, but as the time passed the expedition became the butt of every would-be critic, caricaturist and lampoonist. Saturday 5 November issue of the *West Australian* displayed the headlines: 'A thousand birds in luck—Machine gun jams.'

'Rip Van Winkle' (Vic Riseley), a keen satirist, in his weekly *Sunday Times* 'Smile Again Club' had a Roman holiday. His column still makes interesting reading if one can identify the local characters, on each of whom was bestowed a humorous corruption of their name and title. Thus on 25 December 1932 the Smile Again Club featured *inter alia*: 'Father Christmas Presents Gift to Members: To Major Mereditch of the Royal Emusiliers, I have pleasure in presenting a feather duster of emu feathers, presented on behalf of the wheat growers of Campion.'

The final straw was the suggestion that the earnest, persevering, very English major who, during the campaign, had learned to eat and like emu steaks, should be recommended for the O.B.E. 'for services beyond the call of duty' (one bloody emu).

Following harvesting of whatever crop was left, the majority of the surviving enemy re-migrated to distant and less perilous climes. Thereafter, individual control, plus a bounty of 2s 6d per head, kept their numbers at manageable levels.

In one of the neighbourhood rallies to free their isolated district of stragglers, the enthusiastic Charlie Mann of Yanneymooning, 25 miles north of Campion, organized a drive for the Sunday following cessation of hostilities and the departure of the troops for the south. The plan was to assemble every available man to drive the emus along the Fence into a trap which had already been constructed.

The trap was filled with little difficulty and several volunteers entered the yard with pick handles, to break the birds' long necks. They managed to kill about twenty-five, but gave up the idea after the terrified emus had carried away the temporary fence, and one farmer had had two ribs broken by the over-enthusiastic pick-handle work of a neighbour.

Charlie Mann then suggested that as emus now, understandably, took fright at the appearance of any man, they should catch a large cock bird and dress him up in men's clothes. Thoroughly alarmed and confused, the cock would seek the comfort and security of his own kind who, in turn, would not have a bar of him, and so would set in motion a sort of perpetual chase.

Charlie volunteered to catch the emu with a lassoo made from a set of cotton reins. Standing in the back of a utility doing at least 35 miles an hour, they soon came level with the bird of their choice. Charlie threw, but the noose landed over a fence-post instead of the emu, and Charlie was jerked out of the back of the utility.

Multiple bruises did nothing to check his indomitable spirit. The cock was chased into another hastily erected yard where twenty of the local farmers wrestled him to the ground, and he was fitted with a white shirt and a pair of trousers with the legs cut three-quarter length, back to front. The wives' contribution was to stitch these two garments on with binder twine, whereupon a lady's stocking was drawn over the emu's head and neck.

Let go, he took off like the wind. Local legend has it that the last report received was from a mystified prospector, who swore he saw a 'half man, half emu' chasing a large mob of emus over the South Australian border.

Following the exodus over the years 1930-31 of newly settled farmers, grasshoppers had gradually developed from nuisance to plague proportions by reason of their being able to breed undisturbed throughout six years of perfect conditions, and without hindrance from the normal controls of ploughing and cropping on the arable land of abandoned properties.[6]

J. B. (Jack) Curlewis, field inspector attached to the Merredin branch in 1937, recalls his first experience with grasshoppers in plague form with a revulsion undimmed by the years:

> I was driving north on the Walgoolan Road one day and about three in the afternoon came opposite Wally Wall's place; the road was literally a moving carpet of grasshoppers. When they were crushed by the wheels of the car, they made the surface so slippery that I

could only go at snail-pace, and that lasted for 3 miles. Near his house Wally had 25 acres of wheaten hay some three feet high and just ready to cut—and in those days when horse-transport was still pretty general in the country, hay was still a very important commodity. When I came back that way about the middle of next morning, there weren't any hoppers on the road. They'd moved onto the hay paddock, and had completely eaten it out.

The ordinary farmer could neither withstand the onset of the grasshoppers nor afford the immediate control measure of poisoning. When property after property was abandoned, however, the government instituted control measures under the direction and supervision of the Agricultural Bank's country branches and field inspectors. All cleared, abandoned country within the affected areas was ploughed—not only to expose and destroy the existing egg-beds, but to create the condition of loose soil unfavourable to the formation of further beds. As a corollary to the control measures, many a struggling farmer found a useful form of 'off-farm' income in such ploughing, at the initial rate of 5s per acre on abandoned properties, and on occupied properties at 2s 6d an acre for an approved area in excess of normal fallowing: at the time, the crop area was restricted to 250 acres.

All possible avenues of grasshopper control were investigated, even to the claims of a flamethrower. Whilst the farmers extolled its virtues, the Agricultural Bank and Department of Agriculture refused participation. The flamethrower, fueled with a kerosene/dieseline mixture, was considered costly, cumbersome and dangerous[7] and not practical on dry pasture areas. By the time the grasshoppers had been destroyed, the feed would also have been burnt and the farmer, no better off, was out his time and cost of fuel. There was also the theory that the farmer would, in the long run, lose less wheat by letting the emus graze quietly and perhaps wander on to someone else's crop rather than repeatedly frightening them off. Veritably cases where 'you couldn't win'.

Despite the continuing application of control measures, it was to be many years before the plague was to become merely history, and an item of state history which few Western Australians now know of or recall. The last report in plague form which I have any knowledge of occurred as late as 1948, and I learned of it only recently from Fred Glaskin, who was manager at the Rural and Industries Bank of W.A., Kununoppin branch, in that year.

'I was driving to Lake Brown with a passenger, when I stopped to have a look at a crop', Fred related. 'I happened to glance back at the car, and my passenger was clutching at the bonnet and swaying in a most peculiar fashion.

' "What's the matter?" I called out.

' "I don't know", he replied feebly. "Either I'm car-sick or terribly giddy. The whole road seems to be moving."

'What was happening, and I was accustomed to it, was that the gravel surface of the road was covered with young, newly hatched hoppers on the move. Their colour matched the gravel, and it looked as if someone was pulling a brown velvet carpet in front of the car.'

Although such lingering manifestations of the plague lasted for many years in isolated pockets, generally the controls exercised over some four to five years relieved the situation. With wheat increasing in price again, it was possible to lease the hopper-treated tracts for cropping at $1\frac{1}{2}$ bushels per acre plus the cost of ploughing—which gave the state some return against its expenditures on control.

A number of factors other than the drought, emus, and grass-hoppers—economic conditions, recruitments to the armed forces, and moves to the city for defence employment among them—resulted in the abandoning of still more farms. Farm horses were left to their own devices; fences were few and far between, and with a shortage of both water and feed the wide-roving mobs became a constant nuisance to the remaining farmers.

Head Office instructed Mitchell to organize the collecting of these horses, which represented part of the Bank's assets, and to hold them to best advantage. The instruction followed the usual form: an order with the minimum of direction. Its execution was at the discretion and ingenuity of the branch manager, who was further hampered by advice that expenditure thereon was to be minimal.

It so happened that adjoining the Rabbit Proof Fence east of Maya there was an abandoned farm with adequate paddocks with enough grass and water to serve as a holding ground over the summer months. Mitchell had scarcely made his dispositions about the horses when Ida, whose health had been failing for some time, became ill. He took her to Nancy's place at Tinkurrin. She appeared to recover, and he left for Koorda.

The drive was well in hand. Three stockmen, Ken Clausen, Bob Moralee and Wally Caporn, local farmers who owned riding hacks and saddles, had been recruited. The drive was two-pronged—first they were to muster some eighty horses running loose as far east as Westonia, and bring them to holding paddocks at Newcarlbeon, about half-way between Koorda and Kulja. Then they were to receive, at Koorda, ninety-two horses railed from Warralakin and Boodarockin in order to relieve the farmer-owners of the expense of hand-feeding them until next seeding time.

At this juncture Ida had a second relapse and Nancy wired her father to come at once. On Melbourne Cup Day 1940 Mitchell arrived in Perth with Nancy and his wife and they put up at the United Services Hotel. There was only one doctor as far as the Mitchells were concerned, their old family doctor and firm friend, Dr B. C. Cohen, who had been stationed in Narrogin from 1920-25, first as partner with Dr D. W. Mackie and later in practice on his own. Ida was immediately ordered into the Mount Hospital for observation, but on her insistence, and with the doctor's permission, Mitchell returned to Koorda. Nancy stayed on, living with friends.

All had gone well with Mitchell's horse-catching undertaking, but two days before the start of the final drive he received advice that Ida needed surgery, and that his permission was required. She had begged the doctor to proceed without worrying her Charles, 'who had a lot on his mind', but neither the doctor nor Nancy would take the responsibility.

Arrangements had been made for Mitchell to follow by car on the final drive of some 75 miles to Maya, pulling provisions and camping equipment on a trailer fabricated from the rear axle and wheels of a Peacock Chevrolet car, and fitted with the seat and back-end from a Ford Model T utility. With Mitchell unavailable gimlet shafts were fitted to the trailer, and a quiet old mare selected to pull it, with one of the stockmen taking turns at the reins. The morning the drive left Koorda was a near tragedy. The horses—rested, watered and fed, and exhilarated by the crisp morning—decided to panic. They stampeded in the direction of Newcarlbeon and the three stockmen set after them. Frank Collett, the man in charge, finished his breakfast at the hotel and proceed leisurely to follow the drive by car.

For some miles out, the mallee and tamar scrub formed an almost impenetrable wall on each side of the road, and at one stage Frank was intrigued by the odds and ends of food, camp gear and blankets strewn about, until through the scrub wall there appeared a number of very scared Aborigines, only slowly regaining their normal colour. They had been camped on the edge of the road, sleeping soundly, when they were awakened by the Wild Western sound effects of the stampeding horses thundering among them. Frank saw only the comedy of it all.

The horses were delivered to Maya on 17 December, seventeen days in all from start to finish.

The ninety-two horses belonging to farmers were returned by road to Koorda in March 1941, and railed to their various destinations east of Campion. Mitchell was there to tally them and see them on their way, after which he was free to return to Perth and his wife. It had been found necessary to remove one of Ida's kidneys: her recovery was slow and her convalescence possibly extended by the sailing of Bricky for overseas[8] and the imminent departure of Ginger, who had joined the R.A.A.F.

The family returned to Geraldton as soon as Ida could travel, but at home she still required nursing and constant attention. Mitchell, who had accumulated ninety days arrears of annual leave, decided to apply for the total amount due in order to be at home full time to lighten Joan's load. Nancy subsequently came to Geraldton with her two babies, to live with the family and share in the work of nursing her mother.

In early March Frank Collett, as branch manager at Kununoppin, applied for three months long service leave to enable him to visit Melbourne, and there 'move Heaven and earth' to get into the Air Force. Head Office advised that the leave could not be granted as no relief manager was available. Frank was a great favourite with Mitchell, with whom the long inactivity of leave had begun to drag. He felt he was in the way at home, with both daughters living in, and on hearing of Collett's predicament he volunteered to return to duty. For three months from 17 March he relieved at Kununoppin. Frank and his wife offered him the full use of their house, but the independent old 'cuss' insisted on batching, preparing his meals in the washhouse and sleeping on the side verandah. Among the townspeople was a Mrs Hewitt, the widow of a local farmer, a kind-hearted lady who, as a young girl, had known Mitchell in Narrogin. She thought it terrible that a nice old man should live under such conditions, and she did her best to persuade Mitchell at least to let her prepare his meals and do his washing. Naturally his intimates kidded him about this feminine concern for his welfare, and to such an extent that by the end of his term, so the story goes, he took to the bush every time the good lady came in sight!

T. F. (Tom) Jones, Branch accountant, enjoyed Mitchell's term of office; it was his first acquaintance with this extremely practical manager.

Early in the piece, Tom handed Mitchell the file of a Beacon farmer about whose wheat deliveries recent Head Office correspondence had complained bitterly: it seems that some sort of manipulation had caused his harvest to appear far smaller than the district average in relation to the area he had sown, and had left the Bank lamenting the non-appearance of its anticipated share.

Mitchell studied the file, called for Jones, and said: 'This needs an on-the-spot investigation. You know this bloke, so you'd better come along with me.'

While they were driving along the road to the Beacon area there was time to talk, and Mitchell—as was his wont—began to philosophize.

'You know, Jones, you must never lose your temper with a farmer client. It's not fair to him, and it upsets your own judgement. I've often got into trouble over my temper, and now I do all I can to keep a rein on my tongue. Especially when I have to tackle chaps like this. Never be acrimonious . . .' and so on.

Tom was intrigued by the confidence, and for a while he sat thinking it over. One vital point occurred to him.

'Just how are you going to approach this bloke, Mr Mitchell?' he enquired.

'Well, Jones, it's never any use beating about the bush', Mitchell replied. 'I'll just say to him: "As far as I can make out from your file, you're just a bloody thief!" '

Unfortunately the recalcitrant farmer was not at home on that occasion, and when Mitchell finally caught up with him, Tom Jones wasn't a member of the party; so it's impossible to record, here, the result of Mitchell's 'non-acrimonious' approach.

Mitchell's son-in-law Colin was called into the Army in December 1941. Joan, his wife, stayed on in Geraldton until the fall of Singapore in February 1942 when, with the threat of possible invasion, she, like many others who moved inland, left to stay with H. W. Biglin and his family at Kondinin. At the same time Ida went to stay with Nancy at Tinkurrin. She rejoined Joan and Peggy with their families in Narrogin when they were able to regain occupancy of the family home in Furnival Street in 1942.

In September 1942 word was received that 406812 Flight Sergeant Philip Rossiter (Ginger) Mitchell, as member of the crew of a Wellington bomber, had been reported missing in a daylight raid on Kiel.[9]

For some time Ida's memory had been failing, and now there was a further deterioration. Neither then or at any time in the future would she ever admit that Ginger's absence was anything but temporary. Mitchell was hard hit, and according to the family he took the news very badly. For the first time in his married life he did not have Ida to lean on. No longer was her sustaining influence, her love, her understanding, readily available to him. He had never been selfish, he had only been singularly fortunate in the choice of a wife. For her part Ida needed her husband: she required his presence, needed to know he was never far away. To add to Mitchell's mental turmoil he had been appointed, on the entry of Japan into the war, the Regional Authority through whom must pass all applications from farmers and farmers' sons for release to join the Armed Forces. However, farmers found that it was nigh impossible to get his recommendation for their release. Mitchell remembered the blithe abandon with which he had gone to war, he recalled the futility of it and the subsequent frustration of being manpowered during World War I. This appointment went against his grain; he would have preferred to have no part of it. Who was he to sit in judgement?

Finally, Mitchell could stand the separation and worry no longer. He applied, on compassionate grounds, to be retransferred to Narrogin, and with the subsequent posting of Robin R. Walters as branch manager to Lake Grace, the vacancy was granted to him on 8 March 1943. His main responsibility in his new post was to keep the managerial seat warm. There was practically no new development, no farms were changing hands and wheat and wool were being pooled for guaranteed dividends. He was able to be at home most days and nights.

On 20 July 1944, having attained the age of sixty-five years, he retired. The event is recorded by the *Narrogin Observer* (22 July 1944):

> On Wednesday last, Mr. C. D. Mitchell, local manager of the Agricultural Bank, was the guest of honour at a dinner arranged on his behalf by the local staff of the Bank. Mr. Mitchell, who has reached the retiring age, joined the staff of the Bank in 1908 and with the exception of five years, has spent the whole of the time in Narrogin. Mr. S. B. Donovan, Chairman of Commissioners of the Agricultural

> Bank, accompanied by Mr. W. H. Cover, an old time resident of Narrogin, came from Perth especially to be present at the function. Mr. Donovan, on behalf of the whole of the Bank's staff throughout the state, made a presentation and expressed the regret that was felt in the retirement of Mr. Mitchell who ceased duty yesterday.

He was paid an allowance of £445 19s 8d, representing accumulated annual leave (not taken) plus proportionate long service leave which, for payment of superannuation, ended on 15 April 1945. He had taken the maximum superannuation units allowed for his salary rate, ten in number, worth £2 10s 0d p.w. at that time. At his death they were worth £9 5s 0d, little enough for a lifetime of devoted service.

This and preceding chapters have tended to be critical of the lack of foresight in the development of the eastern and north-eastern areas of our wheatbelt. This criticism was intended but never in a destructive sense. My own thoughts are well expressed in a quote from Margaret McAleer's 'Some aspects of the development of the Northern Wheat Belt':

> The 1931 disaster did not invalidate the development of the Wheat Belt, nor the work of its chief artisan Sir James Mitchell—the man who saw what needed to be done and did it—but it sharply drew its geographic limits and bored into the unsound parts of its economic structure: a necessary and healthy process if it could have been effected less drastically and voluntarily.[10]

Mitchell retired exhausted and possibly dispirited, but his efforts had not been in vain. His work has focused on closer settlement activities before World War I, and on soldier settlement and debt adjustment after it. During the depression years he had shifted to Marginal-area reconstruction, and during World War II had administered other special assistance schemes, including the proliferation of undertakings for subsidies, price guarantees, marketing schemes and others. The fact that these had even been attempted had irrevocably committed government participation through Australia—wittingly or unwittingly, willingly or unwillingly, both widely and deeply—in influencing the basic features of land settlement and rural finance. The majority of the schemes of assistance to farmers were no longer following the old pattern of political expediency, but were tending more and more to be realistic and practical.

Increasing involvement in the formation of policy relating to initial farm size, production programmes, priority of development and settler selection committed governments heavily in the settlement of Crown lands. As a result it became inevitable that they should continue contributing in a patriarchal but modified form toward the attainment of what has now evolved as a relatively open market for farm credit.

When the end of World War II was in sight, and victory seemed assured, Western Australia was fortunate in once more having at hand a 'man of the moment': the Honourable Frank Joseph Scott Wise, Minister for Lands in the Willcock government from 1939 to 1945.

As 'ex officio' Minister in Charge of the Agricultural Bank, Wise had several consultations with the then Premier, the Honourable J. C. Willcock, as far back as 1940, pointing out that a re-organization of rural credit facilities in Western Australia was overdue and that steps should be taken to change the Agricultural Bank from its existing status and activity to a general trading bank. At that time the trading banks were taking over from the Agricultural Bank, every year, many accounts in which they had had no interest at all until the Agricultural Bank had nurtured and nursed them to a sound and healthy state. Millions of pounds worth of sound accounts were lost to the Agricultural Bank because the Bank had no trading facilities for customers. Wise had pointed out that now, almost fifty years after the Agricultural Bank was established, there were three courses open. The bank could be liquidated, at the same time salvaging as much as possible from doubtful accounts, or the private trading banks could be invited to take over all accounts attractive to them. He himself could not possibly subscribe to either of these negative steps. Rather did he recommend a state trading bank with a constitution similar to that of the Rural Bank of New South Wales, aimed at achieving in Western Australia something 'useful to the farmer and at the same time fair to other sections of the community'.

The creation of such an institution was first mooted by Premier Phil Collier's Labor government as far back as 1929. His Excellency the Lieutenant Governor, Sir Robert Faze McMillan, in delivering his address on the opening of parliament on 25 July stated, *inter alia,* under 'Legislation': 'You will be asked to give consideration to a measure dealing with the establishment of a Rural Bank . . .' The idea, however, was stillborn. There is neither reference to further discussion nor the introduction of legislation during that parliamentary session. The concept represented advanced thinking, and with the then recognizable decline in Western Australian economy it may have been considered desirable to defer for a more propitious atmosphere. Somewhat the same condition still existed. Premier Willcock, while sympathetic, advised his colleague to bide his time. Failure was not to be risked.

With the consent of the Premier, Wise accepted the chairmanship of the Rural Reconstruction Commission, appointed by the Honourable J. B. Chifley as Minister of State for Post-War Reconstruction in February 1943.[11]

The further Wise progressed in the investigation and development of the overall report, the more he became convinced of the necessity and urgency for the progression of the plan he had discussed with Cabinet in 1940. His philosophy was crystallized with the views expressed by the Commission in Chapter 5, Clause 703, of Report No. 4, 'Financial and Economic Reconstruction of Farms':

> Consistent with the expressed view of the Commission, Government sponsored financial relations with settlers would be administered through one organisation in each State.
>
> Such an organisation would take the form of a rural bank which would operate in two departments; one providing normal rural banking facilities, and the other acting as the agent for the Government in—
>
> (*a*) collecting existing Government loans and land dues, and
> (*b*) operating an administrative and financial service for the reconstruction of unsatisfactory settlement.

With the completion of this particular report Wise considered it apposite to relieve Secretary Brownlie from his Commission duties, instructing him to spend some time in New South Wales studying at first hand the structure and workings of the Rural Bank of New South Wales. Having spent some weeks in close observation and questioning, Brownlie returned to Western Australia to work on the preparation of the first draft of a Bill to remodel the Bank. Armed with this draft and with his arguments strengthened by experience, Wise persuaded Cabinet, late in the 1944 session of parliament, to make vital changes in the structure of the institution he had so persistently advocated.

The Rural and Industries Bank Act, No. 51 of 1944, was assented to on 1 February 1945. The Bank, in its new guise, opened for business on 1 October of the same year. There was no blare of trumpets at the opening—only well-ordered confusion, plus a certain amount of trepidation.

The spirit of the Agricultural Bank was preserved by the creation of a Government Agency Department within the new Bank 'to perform and exercise such duties, functions and powers as may be delegated to the Bank as the administrative authority from time to time'.

Subsequent governments, whether politically conservative or otherwise, influenced partly by demand and partly by precedents that have proved their worth, have not been ungenerous in implementing the funds of the Agency Department. But no longer is there the sweeping act and its implied limitless liabilities. Separate delegations now provide for assistance not only to certain primary industries but also, at times, to specific areas. The Bank, as administrative authority, has been given full authority and discretion. Personality and previous farming assistance, and provision of a personal equity, are prerequisites to assistance. For the proven trier, a sophisticated farm management advisory service is available free of charge. Political interference is unknown.

It would be of interest here to make a short comparison between the old and the new institutions involved, and to list some aspects of the change-over. The Report of the Agricultural Bank Royal Commission, dated 3 October 1933, gave details of advances made since the inception of each institution named to 30 June 1933. These totalled:

Agricultural Bank	£9,038,338
Soldiers Settlement Scheme	5,990,016
Group Settlement	7,173,764
Industries Assistance Board	13,049,593
	£35,251,711

This must be judged a not insignificant contribution to the development of Western Australia's agricultural resources.

It was estimated by the Farmers' Disabilities Commission, appointed in 1930 to 'enquire into the disabilities affecting the agricultural Industry of Western Australia', that the 20,600 wheat farmers in the state owed more than £32m or five times the value of that year's wheat crop. Of this amount £14m was owing to the Agricultural Bank by 12,000 of its clients. About £11m was owing to the private trading banks and more than £7m was owing to merchants, machinery and superphosphate firms, stock agents and country storekeepers.

The assets of the Agricultural Bank, representing the balances of outstanding loans (buildings were not owned), reached a peak of £16,523,510 at 30 June 1935. This figure included the 'Sundry Borrowers' under the Group Settlement Scheme, which had been taken over by the Agricultural Bank as from 1 October 1930.

By 30 June 1945, the latest balance date before conversion to the Rural and Industries Bank, this total had fallen to £9,264,687, more from write-downs and adjustment of debts than from repayments of principal by borrowers or discharge of mortgages by the private banks. By September 1952 the greater part of this remainder of £9m had been absorbed by the Rural and Industries Bank as capital.

The most staff ever employed by the Agricultural Bank was 266 at 30 June 1932. There were eleven active branches, including the Perth office, operative at 30 September 1945. On the opening of the Rural and Industries Bank, the staff totalled 227, of whom less than 6 per cent were officers with previous trading bank experience recruited for the purpose of instructing and guiding the original staff on the methods and techniques of trading bank practice. How well this 94 per cent had been trained by Mitchell and his contemporaries in human- and customer-relations and imbued with dedication and purpose is revealed in the progress of the new bank.

Rolland William Bosisto, full of confidence and with a judgement backed by years of experience, resigned from the position of manager, Branch Department, National Bank of Australasia, Perth, to become the new Bank's first chairman of commissioners. His leadership characterized by a supreme optimism, courage, humanity, and a minimum of interference, were basic factors in achieving the seemingly impossible. He was assisted by Board members Cyril Abey and Ross Wallace Brownlie, promoted from general manager and secretary, respectively, of the old Agricultural Bank. In agricultural reconstruction, there was no more experienced authority in Western Australia—in Australia, for that matter—than Brownlie.

Skilfully led by the upper echelons, the staff moved into financial fields hitherto unknown. Ridiculed and scorned by certain banking rivals, and reminded at every opportunity of the short life of the Primary Producers Bank[12] (which was forecast as their fate) they retaliated with greater effort. Eventually the pundits, who facetiously referred to the new bank as the 'Cockies' Bank', were quietened with this challenge: 'There's a gold medal as big as a soup plate for the first to prove that a trading bank moved into any wheat or sheep farming area ahead of the Agricultural Bank.'

At 31 March 1970 the Rural and Industries Bank, heir and assign of the old despised and too-often ridiculed Agricultural Bank, had assets totalling $195,359,763, including loans and bills discounted of $97,370,002, and was holding customers deposits of $82,222,825 in the General Banking Division, and $68,815,989 in the Savings Banks. Branches totalled fifty-seven and staff 803. In addition, the Agency Department of the Bank administered on behalf of the Commonwealth and state governments, loans to farming and other industries totalling $13,041,242. The seeds that James and Charles Mitchell helped plant have come into full bloom and yielded a hundred-fold.

1. Constance Mabel, daughter of Maitland Brown and Amy Frances Howard, married Walter McKenzie Grant on 31 March 1881. Maitland Brown sold 'Newmarracarra' station to his son-in-law, Walter, on 23 October 1883. His son Walter married Peggy Mitchell on 8 October 1940. There were three children from the marriage: Rossiter McKenzie, Donald McKenzie and Helen Joan McKenzie.
2. The Wheat Industry Assistance Act No. 53 of 1938 provided £2,000,000 for the elimination of Marginal wheat areas. Funds were allocated to:

New South Wales	£800,000
Victoria	177,000
South Australia	713,000
Western Australia	417,000
	£2,107,000

The additional £107,000 was provided from Commonwealth funds in 1941-42 enabling the states to carry out their full programme for that year. The final plans of the states showed a total cost of £2,950,000, comprised of:

New South Wales	£1,620,000
Victoria	200,000
South Australia	713,000
Western Australia	417,000
	£2,950,000

In 1946 an additional £843,000 was appropriated from Flour Tax, which amount otherwise would have been payable to No. 9 Wheat Pool (Wheat Industry Assistance Act No. 71 of 1946, amendment to No. 53 of 1938). The total appropriation for elimination of Marginal wheat areas appears to have been:

Wheat Industry Assistance Act	£2,000,000
Commonwealth Funds 1941-42	107,000
1946 Flour Tax	843,000
	£2,950,000

Funds provided were titled 'Commonwealth Grant Assistance Distressed Wheat Farmers'. Employment of the funds in Western Australia resulted in the whole or partial reconstruction of 1,053 farm holdings. Total expenditure amounted to £301,250; the unexpected balance was transferred eventually to the Wheat Industries Research Fund.

3. WHEAT ACREAGE RESTRICTION

Area cleared (acres)	Area for wheat for sale (acres)	Balance to complete 5-yr rotation stock, feed and conservation (acres)	Total area cropped (a) (acres)
Up to			
2,000	250	150	400
2,250	280	170	450
2,500	310	190	500
2,750	340	210	550
3,000	375	225	600

(The increase (a) is proportionate to the area cleared.)

This cropping rotation is concerned also with efforts to eradicate grasshoppers, and to re-establish barley grass, especially in the north-eastern district, as it seems this valuable feed plant is disappearing as a result of overstocking, grasshoppers, army grubs, and web-worm. (Extract from the records of the Rural and Industries Bank of Western Australia, dated 13 April 1949.)

4. He was born 4 May 1885, son of T. Hope Murray, of Beevor, South Australia, who was the second son of the original John Murray, founder of the Murray Merino breed of sheep. Following his retirement, he lived at 87 Princess Road, Claremont. He died on 16 February 1958, survived by a widow and one daughter.
5. Daniel O'Leary, farmer of Walgoolan (former General Secretary of the Wheat Growers' Union), had been prominent in the representations for the machine-guns being brought into the area. Apparently part of the original arrangements was that the cost of the ammunition would be found by support from various firms and subscriptions from farmers, and also that the representatives of the armed forces would be both accommodated and have transport provided during the campaign.

'After the abortive campaign O'Leary was asked to pay £24 to meet the outlay of ammunition. Being a man of varied resources, he promptly put in a counter-claim saying that he was prepared to overlook such minor charges as rent for trenches and recompense for cleaning up the battlefields after engagements, but that he expected immediate settlement of the following account:

Victualling His Majesty's troops	£9. 0. 0
Transport of troops	10. 0. 0
Damage of transport vehicles during conveyance of troops to front line	5. 0. 0
6,000 bushels of wheat @ 3/- a bushel (eaten by enemy during official delay in opening counter-offensive)	900. 0. 0
	£924. 0. 0

It is understood that all the emu war debts were finally written off, as were many other debts relating to Eastern Wheatbelt after the grim days of the depression.' (C. F. H. Jenkins, 'Birds proved a tough enemy in 1932 Emu War', *West Australian* 12 August 1963, p. 5.)

6. The term locust is applied to those types of grasshoppers which, under favourable seasonal conditions, breed in large numbers and mass together in swarms, and which may undertake long migratory flights. The Western Australian species to which we refer is known as the Little Plague Grasshopper (*Austroicetes cruciata*). It is not a true locust and is often referred to as a gregarious grasshopper. (C. F. H. Jenkins, 'The Plague of Grasshoppers', *Journal of Department of Agriculture, W.A.*, no. 4., December 1937, pp. 367-80.)
7. A similar type of flamethrower, being tested in the Salmon Gums district for control of mallee suckers in August 1947, exploded and seriously injured M. L. Austin, chief valuer of the Rural and Industries Bank.
8. Corporal Charles Rossiter WX 7603 (Bricky) sailed from Fremantle on 3 January 1941 as a member of the original 2nd/28th Battalion, 'D' Company. Seriously wounded at Tobruk on 3 August 1941, he was repatriated to Australia on 7 December 1941. He married Hazel Bell on 22 July 1942 and had four children: Kimberly Charles, Albany Philip, William Murray and Elizabeth Anne.
9. From official records: 'Flight Sergeant Mitchell was attached to the Royal Air Force United Kingdom and under the command of Pilot Officer J. S. Ward left the R.A.F. Station, Wickenby, in a Wellington 3 JB 776 at 7.30 a.m. on 26th September, 1942, on a sea-mining mission in the Kiel Bay area. No further news was heard of the aircraft or crew after take-off.'

10. *Journal and Proceedings of the Western Australian Historical Society*, vol. 5, part 2 (1956), 45-70.
11. It was rather appropriate that Frank Wise, appointed as Premier, Treasurer and Minister for Lands and Agriculture on 31 July 1945, should open the institution he had sponsored.
12. The Primary Producers Bank was the only new banking company to be formed this century. It was promoted in February 1923 in Queensland, and closed its doors on 24 August 1931. It was a bank for the man on the land, and this was its weakness. In confining its loans to farmers it depended too heavily on a section of the Australian economy which was brittle in depression. Another weakness was its youth: it had to chase business where other banks were well established. It had to lend generously and build premises at the expense of its profits.

As prices of primary products fell, so the bank reserves of cash were steadily withdrawn. J. F. Brett, deputy chairman, realizing that there was no hope of increasing cash reserves, approached the National Bank, handed over confidential files and suggested a merger. No merger resulted. Aid came from the Commonwealth Bank in the form of £145,000 but in the eighteen months prior to August 1931 it lost 40% of its deposits. Rumours caused a constant drain on the bank. The Commonwealth Bank asked all trading banks whether they would assist, or let it go into liquidation. The Primary Producers Bank of Australia closed its doors on 24 August 1931. They paid 10s in the pound to liquidators.

Eight branch offices were established in Western Australia: Perth, Katanning, Geraldton, Kojonup, Northam, Goomalling, Merredin and Trayning.

19

The Long Afternoon

Retirement did not bring Mitchell back into Narrogin's community life. There was no zest to it now. Eight years had seen many changes; friends in high places had been replaced by younger men, mostly strangers. His wife was always happier when her Charles was near, and he spent all the time he could in her company. With at least a third of an acre of spare land at his disposal he became a keen vegetable gardener, but refused to grow or tend any flowers other than geraniums. *Yates Garden Guide* became his bible, and he could recite most of the vegetable section by heart. The garden was not designed for one home's requirements. Typical enough of the man who did not know the meaning of half-measures, the whole of the land had to be tilled and the neighbourhood benefited accordingly.

However, his immense knowledge and experience could not lie entirely fallow. Pressed by the local solicitors and stock firms he consented to carry out farm valuations for probate or sale purposes, and there was no-one more skilled or conscientious for the job. His sketches of improvements which accompanied each valuation were models of neatness, with the soil types classified and coloured blue, red, white or yellow—first, second, third class and salt-affected respectively. An integral part of his equipment was a boy's schoolbag slung over the shoulder and carried on his back, containing field book, sketches, compass, ruler, protractor and refreshments. In later years he carried a peeled gimlet stake 'just to help me over the rough patches'. The Taxation Department offered him full-time employment (Exempt Valuer, Grade I Temporary) based at Narrogin, but he had had enough of regimentation; at last he was free. His casual valuing and small pension brought in sufficient for their modest needs.

His valuation reports needed to be typed, which created the opportunity he so wanted: to renew, with reasonable excuse, an association with members of his erstwhile staff.

As the mooted transfer of the Agricultural Bank to the Rural and Industries Bank of Western Australia developed in the committee rooms of state parliament, office work within the Bank became minimal. The ever-thoughtful, efficient, generous-hearted Thelma Flute was still there with time to spare, to set out and type her old boss's reports.

Valerie Elliott, who joined the Bank at Narrogin in January 1945, aged sixteen, recalls very vividly her impressions of Mitchell: 'a big man with a hearty laugh, who invariably arrived round afternoon-tea time and stayed on yarning away till knock-off time'.

Once started, Mitchell held the flow of conversation. Standing with his back to an old mantelshelf, cup of tea in hand—getting colder and colder—he held Valerie entranced. She was initiated into a world whose existence she had never suspected, an exciting world to which she was later to contribute as the wife of Vernon Bannon, a bank officer and subsequent branch manager of the Rural and Industries Bank. She recalls, for instance, a mid-winter afternoon when, with the manager absent on business, they closed the doors at 3.30 p.m., and the entire staff sat around an open fire and yarned. And on another occasion when they were short of milk, Mitchell took the jug and milked sufficient from one of the neighbour's cows which were allowed to graze in the Bank's yard.

These days were not to last. The Narrogin branch of the Rural and Industries Bank opened for business on 1 October 1945,[1] and in the ensuing frenetic haste, Mitchell's story-telling visits tapered away. The staff had began to talk a different language, new faces began to appear. He was lost and felt no part of the rapidly changing atmosphere of the office he had ruled for ten years. It would have been fitting to write that there was an opening ceremony, with Mitchell present as a guest and receiving due acknowledgement, but it was not to be.

For the time being his days were happily filled by the closeness of his daughters Joan and Peggy, and their respective children, who lived at the Narrogin home while their husbands were in uniform. Bricky had recovered fully from his wounds, was married and was working with the Repatriation Department. Tobacco was still rationed but Mitchell had enough friends to fill his needs with 'Dark Havelock plug' from 'under the counter'.

They were soon split up, however. Joan and family joined Colin when, on discharge from the Army, he was reappointed branch manager of Elder Smith and Co. at Geraldton; and in 1946 Peggy and family accompanied Walter to Yorkrakine, where, while awaiting allocation of his own place,[2] he managed farms for the War Service Land Settlement Scheme.

Coincidentally, their places were filled by the return of Joan and her children. Colin had been promoted to an organizational position which took him all over the state, and with the post-war housing shortage to contend with, the family home at Narrogin presented a happy solution. Colin spent what spare time he could find with them, and Peggy left her eldest son Ross with Joan to commence schooling. This constant company for Ida meant that Mitchell could continue to leave home again, in peace of mind, for his valuing work. Most of all, the jobs he welcomed were those that took him out 'east', where he made the Tinkurrin home of Nancy and Hal his headquarters. Hal looked forward to his visits, which were invariably prolonged by Mitchell's completing and forwarding his reports from their place. Hal, even today, is unstinting in his praise for the advice and help he received. Not only would Mitchell mend fences, make gates, milk morning and night, chop the wood, pick roots, and help with the sheep; he had the gift of never getting in the way. While never overtly advising or criticizing, he was yet always full of helpful suggestions, mentioned casually and always at the right time.

Even in retirement Mitchell could never shake off the bushman habits and routines which had become part of the grain of his soul in the long years roving the back-blocks roads and bush tracks of the nascent wheatbelt areas, camping out in fair weather and foul, or rattling along with his buggy or car-load of hope and good advice among the battling settlers. He could not sleep, which did not please Nancy and Hal, who could. The household would be supplied with a cup of tea by 6.30 a.m. Ten minutes later, if they were still in bed, there'd be a roar: 'Get out of that or you'll get bed sores!'

One of the most touching and characteristic stories of his retirement concerns the foundation of his own personal football team. He had always loved the game and on a Saturday morning in May 1946, attracted possibly by the noise, he walked across the street from his house to view a football match being played by a group of boys on a vacant block adjoining Ted Walliss's place on the west.[3]

The ground was rough and dangerous, and umpiring decisions apparently were decided by whoever yelled the loudest. When he managed a moment of quiet, Mitchell asked how they'd like to play on a better ground, and if they would like him to umpire for them. They liked on both accounts, so the venue was changed immediately to Mitchell's cow paddock, Lot 85. At the conclusion of the match he awarded a shilling each to the best player on each side, chosen by himself. The kids were charmed and delighted. Naturally, they then all sat down and had a talk, the upshot of which Mitchell's block was adopted unanimously as their home ground.[4] Could they bring along more kids and could they play each week? Of course they could, and welcome, but what about an age limit? After some deliberation it was fixed at fourteen, and so 'Us Fellows' Football Club was born, with Mr Charles Mitchell, in his sixty-eighth year, as Patron.

Mitchell coached, coddled, umpired and fed them. Since their quarter-acre 'ground', with plain and barb wire boundary fences, sloped to the rear, 'winning the toss' involved not so much taking advantage of the wind as whether to start playing 'up hill or down'. A clump of manna gums half-way down the western side and a large stump directly opposite it were no more than slight inconveniences to play around. Another stump on the southern boundary became one of the goal posts. Two teams of even strength were selected and, captained respectively by Bert Walliss and Wally Stevens, played each other every Saturday morning. Teams varied from seven to nine a side, according to the number rolling up and clamouring to play. Uniform guernseys were not required, but each boy had his initials sown on the front to his own fancy, with captains and vice captains having the initials of their respective ranks also prominently displayed.

The players began to arrive as soon as they could after persuading their mothers to provide an early breakfast, and the first job was to tether the cow in the adjoining garden and then clean the arena of her droppings. At half-time, Mitch (he had insisted the boys address him as such) served quartered quinces, apples or oranges, and when the match was over, the players, mud and all, trooped into the house to a morning tea served by Mrs Mitchell and Joan.

On the odd occasion when horses, owned by son-in-law Walter Grant, were quartered on the 'football ground', the venue was changed to a paddock owned by the parents of Wally Stevens, behind their home at No. 5 Felspar Street. This 'ground' was not popular, however; besides its being very small, there was an irate neighbour who objected to the ball being kicked over his fence, and who at one time actually confiscated it, only giving it up on the very strong representations of Wally Stevens's mother.

Ted Walliss (father of Bert and Wally, foundation Us Fellows) helped Mitchell with the umpiring and coaching, and the local football clubs supplied the balls. The Patron continued to provide a shilling for the best player on each side, and heaven help the player who showed bad temper or questioned the umpire's decision. Us Fellows adored him, and many of them recall that on their way home from school, Mitch always happened to be outside in the garden, or leaning on the gate smoking his pipe. To a respectful: 'Hello, Mitch!'—he would drop what he was doing and reply to each boy by name. Then, continuing to lean on the gate, or transferring to a seat in the gutter, he would discuss with them next week's teams or tactics, or the next season's prospects. A large seedling apple tree grew in the Mitchell yard and bore prolifically. When the fruit ripened Mitchell always had a few in his pockets for his young friends, and occasionally would excuse himself from their talks with: 'Just a minute and I'll get some apples, you can eat as you talk.'

During the summer the ground became a sports training ground, with a sand pit dug for long and high jumping, and a set of brush hurdles erected for steeplechasing. Mitchell helped cart the sand and teatree saplings for the hurdles in his car, a box-type two-door Chrysler sedan of pre-war vintage, capable of carrying anything from firewood to a football team.[5] Puffing away at his pipe, he attended every practice—participation through observation was his enjoyment.

His example undoubtedly contributed to awakening the local football administrators to the youthful talent available. In any case, a Colts Association was formed in 1947, with Imperials and Towns each organizing a team which met in their first match on 11 May of that year.

When it rained, or they tired, the kids pored through Mitchell's library of Test Cricket memorabilia, usually led in this by the grandson of the house, Michael Wishart. In addition to his passion for football, cricket and athletics, he had inherited Mitchell's interest in horses. He studied performances and, being statistically minded, he became aware of the profits likely to accrue from bookmaking and decided to run a 'book'. Among the first races on which he operated was the Caulfield Cup of 1946, and as soon as the acceptances and prices were published, he hawked his book amongst Us Fellows and his school mates for their penny bets.

On being importuned, Mitchell decided to help out by placing threepence on each horse in the race but, to Michael's dismay, the 100 to 1 outsider, Royal Gem, was the winner. Mitchell insisted that the 'bookie' pay up, irrespective of how long it took him to do so; he must learn to stand by his betting obligations which, to a gentleman, were sacred. Mitchell saw to it, nevertheless, that every now and then Michael's pocket-money was supplemented through his grandmother or his mother, and he constantly regaled his friends at the Narrogin Club with progress reports of how much he'd been paid and how much was still owing.

Their second playing season, in 1947, could not start soon enough for Us Fellows. Furnival Street was their 'headquarters and home ground' of which the *Narrogin Observer* wrote:

> As a football ground, the block has its shortcomings, but it is doubtful whether this worries the youngsters. Prior to last Saturday's homeric battle between seven-a-side teams, representing State and Convent school teams, it was found necessary to drain the block because of a large puddle in the middle of the field. This task was achieved, however, with the same spirit of co-operation and communal endeavour which has marked the brief but successful career of this youthful association.

Growing in size and public interest, Us Fellows attracted helpers and parents, many altruistic, some wanting to climb on the band wagon. Soon there were captains, coaches, presidents, 'lady servers' and committees—in fact, all the necessary figureheads of constitutional procedure—with behind it all the fatherly interest and encouragement of Mitchell. Club rooms were built in the clump of manna gums, using the trunks as uprights. Roof timbers were mallet poles cut on an expedition led by Mitchell in his car. Roofing iron was obtained from the local tip; Mitchell donated the bags and hessian for the walls. The club room served as grandstand and dressing room, and—sticklers for detail—Us Fellows even installed a rubbing-down table.

There was no lack of talent and potential and there were enough Us Fellows attending the Convent to make up a side, led by Mort Kuhlman, with 'Dinny' Barron as his 'aide'.[6] Wally Stevens led the 'home side'. They met each other every Saturday morning—except when Mitchell's paddock held horses—but their 'away-from-home' ground was now no longer in the Stevenses' paddock. Now they played 'away', in the afternoon, on Mr Kuhlman's cow paddock at the rear of the Kuhlman home at No. 79 Clayton Road. Mitchell still supplied the fruit for half-time and morning tea for the home matches. Doctor Jack Savage always came to see his Johnny play, and brought lollies, which were shared enthusiastically.

One particular event in 1947 still lives in the minds of the Us Fellows and supporters. Toward the end of the season a combined team travelled to Williams in Mitchell's, Colin Wishart's and Norm Culleton's cars to play a side of similar age group. The game was played in a paddock opposite today's Mobil garage. No-one can remember who won, but the players still have memories of the very excellent picnic tea to which both sides were treated in the local hall. The parents have vivid memories of the speed and of the relish with which the food was consumed, possibly equalled in history only by Egypt's famed plague of locusts.

The competition had become news, and the *Narrogin Observer*'s photographer arrived to obtain a joint team photo which was published on 15 August 1947. There were two 'young lady' supporters included in the picture, each with her arms folded footballer style. The wise old editor of the *Narrogin Observer,* E. S. (Sid) Hall, sounded a note of warning in an article pertinent to the photo and the activities of Us Fellows, when he wrote:

> Now the local Council wants to help by providing a suitable playing area in the Southern portion of the town. The suggestion is an admirable one, but one wonders whether it is in accordance with child psychology. Environs and sentimental attachment play a big part in the affection of kiddies for some particular mud patch, and these considerations usually outweigh any disadvantages or restrictions associated with a convenient site.

The editor was partly right: Us Fellows never carried on beyond 1947. Some suggested as a cause the number of boys who had moved out of the age group, breaking up the local cliques; but the main reason was probably Mitchell's withdrawal. The family atmosphere so dear to his heart had been lost; too many wanted to share. Instead of adding, they had subtracted.

MITCHELL WAS INCREASINGLY CONCERNED about his wife's declining health, and his time was more and more consumed in caring for her. Colin had been transferred to Head Office early in 1948, with Joan and the family following him to take up residence in Nedlands. Her departure meant that the care of Ida must be shared between Nancy and Peggy at their respective homes:

Mitchell always accompanied his wife, and Narrogin saw less and less of him.

Mitchell's brother, the astute Sir James, had, through successive governments from 1934 onwards, resisted appointment to the status of Governor from Lieutenant Governor. As Lieutenant Governor his appointment was without limit, whereas as Governor he would have had to accept fixed term appointments. There was no government willing to force the issue until—in his eighty-third year, on 5 October 1948—he accepted appointment as Governor for the period ending 30 June 1951.

On 13 October 1949 Clara, his wife and loyal supporter for sixty years, passed away. James's widowed sister, Mrs Nora Rae, moved into Government House as the Governor's hostess. Her duties were not arduous as there was little entertainment—not because of James's advancing age or period of mourning, but because of his paltry salary and allowances. Nora still speaks scathingly of the £2,214 he received yearly as Lieutenant Governor, and of the long delayed increase to £4,000 as Governor.

During James's entire tenure at Government House all members of his family were punctiliously invited to all state functions. Sometimes they would accept, but it was only very rarely that they stayed overnight. Clara was rarely likely to ask them, and they were most unlikely to accept. Neither side had ever completely forgotten that Clara's family, the Spencers, had opposed the marriage—the idea of their daughter's marrying a common bank clerk—and this still rankled among the Mitchells, and it must have been difficult for Clara to back-pedal, even after so long. Actually, she was concerned only with her husband's world—to help him in his work and to raise his children.

Sir James did not long survive her. He stepped down from the governorship on 30 June 1951, and as soon as he had turned over to the Administrator, Hon. Sir Patrick Dwyer, he took his rail coach to live in while he visited his eldest son, Roy, at Donnybrook. He passed peacefully away in the coach, from natural causes, on 26 July. His life's work was done. There was nowhere to go, nothing to live for. A contemporary who knew him well later expressed the conviction that an extension of his term of office would automatically have extended the term of his life.

The state funeral at Karrakatta Cemetery on July 28 was preceded by a service in St George's Cathedral—opposite the same Government House at which Sir James had presided as Lieutenant Governor or Governor for eighteen years, and alongside the Treasury Buildings in which the weight of his personality and experience had been made manifest in many of the state's most vital political decisions. Mourning crowds lined the route to the cemetery, and the then King—George VI—sent a message of condolence.

Charles attended as the sole surviving male member of the 'Willows' family, but found to his surprise and deep hurt that he had not been included in the official mourning party, nor was he mentioned in newspaper lists of the chief mourners. Roy Mitchell, in whose hands had been left the arrangements for his father's funeral, apologized for the curious oversight, but Charles had been deeply hurt, and the hurt remained. Admittedly, he had sent no prior word of his attendance, nor pushed himself forward on arrival. He had visited Government House only infrequently and then usually only if he happened to be in Perth on Saturday to take James to the races. The surviving sisters then resident in Western Australia—Kate, Mabel (Mrs Stephen Fry) and Nora—did not attend. James had held fixed opinions about the suitability of ladies attending funerals, and even in death his wishes were sacred. The other sister then surviving, Marion (Mrs Harry Nankervis), lived in Adelaide.

WITH THE ADVANCING YEARS MITCHELL'S valuing work had gradually tapered off; he had few interests outside of his immediate family and had little inclination to travel. In December 1953 Aubrey Hardie, realizing the despondency that was creeping over his old friend, invited him for a holiday at 'Yundamindera' station some 120 miles north of Kalgoorlie which his family had acquired in 1934. The 500-odd mile trip was made in one day by car, and on that part of the journey between Clackline and Southern Cross Mitchell shed years from his age. Each few miles revived memories, with accompanying inimitable stories. Here were 'Jim's farms', there where he took his bride to their first home. Tales of settling farmers, stories of the depression years and subsequent 'reconstruction' contributed towards the mental pick-me-up he so badly needed. Although it was fifty-five years since he had last been in the Yundamindera area, he observed to Aubrey when driving through the station:

'We're coming on to Pennyweight Flat.'

'Are you sure?' teased Aubrey.

'Of course I'm sure. There should be an old hand pump set into a soak.'

And sure enough, there was.

For old times' sake Mitchell had once again to visit Malcolm, sixty and more years before a hopeful gold producing centre which he had visited as a survey hand, but now another ghost town. No effort now with the car replacing the horse; they stopped to view Cardinia Creek and drive part of the way to Murrin Murrin, retracing in fact his remarkable horse tracking episode of 1897. Aubrey was only too happy to oblige; the 'old chap' was deriving so much pleasure from his trip back along the roads of his early memories.

THE LONG FRUITFUL YEARS OF DOUBLE-HARNESS in marriage with his beloved Ida were now drawing to a close for Mitchell. Her health, which had been indifferent for so long, deteriorated rapidly throughout 1954, and toward the end of the year she was confined to a bed in the Wickepin Hospital. Her death was not unexpected when it came on 12 December, quietly, and with little notice to the family. She was buried in the cemetery in Narrogin with the family and her friends about her.

If this story dwells but little upon her character and qualities it is not because they were not of an outstanding order. She was the perfect help-mate for her ebullient, gad-about, visionary, red-headed husband, and they were the sole loves of each others' lives. She was a staunch back-stop in times of trouble, which Mitchell knew in plenty. She was a careful manager of his home at a time when a few shillings could tip the balance between making a go of it and going under financially. She was an understanding and untiring mother during years when parenthood was not the 'exact science' it is today, but more an instinctive reaction to situations, a remembering of 'what had always been done', a regimen of bedside vigils at night and wise forbearance by day. She was also the good friend of her husband's friends, and reflected onto him the sure and shining light she generated in the peace and harmony of their home.

Picking up the threads of his life without her, Mitchell lived either with Peggy at Katanning or with Joan at Nedlands. He visited 'the shack', as he called their house in Narrogin, at irregular intervals, to do odd jobs and repairs. He refused to let it or sell it.

During the winter of 1956 he contracted pneumonia from driving the Chrysler with a broken wind-screen, and despite his struggles and protestations, was placed in Katanning Hospital. On his first morning the nurse arrived at 7 a.m. with dish and towels, and placed a screen around his bed.

'What the hell are you doing?' he enquired.

Mr Mitchell, I've come to wash you.'

'Like hell you will. I've just had a bath.'

'Mr Michell, don't you realize you have pneumonia?' The nurse was aghast. 'Taking your own bath is very dangerous. You are never to take another bath without permission, understand?'

'Yes', said Mitchell.

Next morning, at the same time, the same nurse arrived with dish and accessories.

'Nurse, you're wasting your time and mine', declared the patient.

'Mr Mitchell, don't dare tell me you disobeyed orders and had another bath.'

'No fear', said Mitchell. 'I had a shower this morning!'

After the doctor and Matron had been called, he got the message, and submitted with a very bad grace, blushing, so his friends averred, each time he received nursing attention.

He could not help but chafe. As he boasted, he had never been in hospital before, even to be born. Throughout his life he had enjoyed near perfect health and besides, there had never been time to be ill. The worst that had happened to him was the nervous breakdown in 1912, when the Hardie brothers had taken him on a month's holiday to their Waralong Station. Pulling one's own tooth with the aid of a pair of pliers and a large drink of brandy; opening an elbow infected with 'Barcoo rot', using a 3-inch nail sterilized in the flame of the camp fire and held with the pliers; drawing the core from a boil with a hot bottle; these were commonplace manifestations of medicine for a bushman of Mitchell's calibre. Apart from a bottle of 'painkiller', the only medicine he carried was chlorodyne for treatment of dysentery contracted from the sometimes doubtful water of holes and dams.

His recovery from pneumonia was slow, and certainly was not helped by his interference in the treatment; he lost appetite, strength and patience, and after two weeks begged Peggy to take him home and let him die in peace. Peggy promptly packed him in the car. To her, Mitchell's requests were commands. She understood how he felt. It was his decision to make, just as it was his own life to live or not. The doctor and nursing staff warned against his leaving the hospital prematurely, but it is doubtful if their protests were anything more than token. They were not displeased to see him leave.

With his sickness and lengthy convalescence, Mitchell had omitted to renew his driving licence, and when he tried to do so, he failed the eyesight test. This did not concern him greatly—the Chrysler was given to one of the Blythe boys who lived opposite their home in Furnival Street. He had tried previously to give the car to grandson Michael. Joan, the mother, refused the gift: she wasn't going to have any son of hers, especially one just turned seventeen, drive such a monstrosity. When he felt like it, Mitchell borrowed the Grants' car, much to their dismay and concern.

As soon as he was able, he resumed his self-appointed occupations. He loved the active farm life, but not the mechanized kind. Putting an old horse into the spring cart, he took off each morning to pick and cart roots; or taking the horse out of the cart and fitting it with snigging harness, he heaped and burned fallen trees and logs. In addition, there was always the bountiful kitchen garden he found time to provide and tend.

A recurrent pneumonia put him back into Katanning Hospital in July 1958, equally to his own disgust and the dismay of the hospital staff. Once again they went through the charade of treatment, and once again he was discharged on his own request and against medical advice: he would never have been described as a model patient.

His recovery was much slower because of a general debility which left him very lame in one leg, and walking required the aid of a stick. Being unable to do jobs he had previously been capable of irked him constantly. He accepted neither old age nor his disability. When visitors enquired after his health he would reply, to the extreme embarrassment of the questioner: 'I'm no good. I'd be better dead.'

He was far from dead, however, and still occasionally stirred up the pool in his home town. As he was never given to boasting, except as a medium for legpulling, his old staff were amazed and delighted to hear in 1958 that he had announced publicly, and with more than usual force, that he had 'trained more senior officers in the Rural and Industries Bank than the rest of [his] contemporaries put together'. The statement was true in fact; at the time his old staff was represented by two members of the

Board of Commissioners, the chief valuer, the senior auditor, and two senior managers. The 'old' staff just had to do him honour and persuaded him to come in and see them, which he did one afternoon. There was an immediate adjournment to the Criterion Hotel. Fortified with a Corio whisky, Mitchell stated: 'I don't know how you fellows can take time off; you all should have had the sack years ago.'

'Oh, it's all right, Mitch', someone replied. 'We've all got rather unique qualifications that give us special privileges.'

'What qualifications have you got that I don't know of?'

'We've now all got M.T. after our names.'

'What's M.T. mean?'

'Mitchell Trained.'

The compliment was well taken. The old chap drew himself up straight, his thin chest swelled and for a minute or two he did not need his stick to lean on. There was silence all round. It was up to him to speak first. 'How's all your families?' he enquired. The conversation was turned but the purpose had been well served.

Four years later, the ever-thoughtful and kindly Thelma Flute (now Mrs Illingworth) arranged to bring Mitchell in to see the Bank's new premises in Barrack Street. His old staff entertained him to afternoon tea and he did his best not to be impressed by the luxury of the appointments and service compared with those of his day, thirty years before. After a tour of the building, he was taken into the Board Room to meet the chairman, George Chessell, who, after remarking on his delight at meeting Mr Mitchell, stated that it was 'the first time I have had the honour of meeting a man who is a legend in his own lifetime'.

Despite these and other diversions, time hung heavily for one whose life had always been crammed with action. Mitchell was restless and querulous; but he brightened towards the end of 1959, when he learned that he was going to be a great-grandfather.[7] For months before the event he trumpeted: 'You'll see the red hair coming out again in the third generation.' But it was not to be; the baby girl was a brunette. Mitchell took this mean trick of Fate to heart but he gradually got over the disappointment, with the often expressed prophecy: 'Oh well, the next one will be redheaded!'

About July 1961 Byron Hardie (Tom Hardie's eldest son) had some business that would take him out of town for some weeks, and as Ruth, his wife, was extremely nervous after dark, he asked Mitchell if he would mind staying at 'Culbina' during his absence. Mitchell, no doubt happy to be once more active and of use to one of his friends, agreed, and left for Narrogin. On his first morning there he took off for Furnival Street to see what wanted doing at the 'ranch'. It was raining, and he noticed the gutter over the back verandah was overflowing. Mounting a ladder, he proceeded to clean the gutter of grape vine leaves. Overreaching, he put too much weight on the old trellis; it broke, and down came Mitchell, pulling the gutter with him. There he was, upside-down, with his legs caught in the rungs of the ladder. With the rain and the overflow from the gutter pelting down on him he hung, drenched through, for some twenty minutes, until rescued by some passers-by. They were attracted not by calls for help, but by his lurid language.

Back again in Narrogin Hospital with double pneumonia, he was too ill to know what was going on, but still resisted nursing attention. Peggy, the member of the family nearest to him, found daily trips from Katanning too time consuming. She took him home to nurse him herself.

It was the beginning of the end. He was now a semi-invalid; even gardening and pottering about took too much out of him. His waking hours were spent with his pipe and his daily nips of whisky, and in talking to his friends. He lost all interest in the ranch, and Bricky—the administrator of his mother's will—disposed of both house and land to the Australian Association (Seventh Day Adventists) in October 1962.

Mitchell had been raised in the grand tradition that 'a family looks after its own', and had subscribed to it all his life, yet now he resented its application to himself. All his active life he had refused to tolerate wasted effort or injustice and now found it difficult to accept the boredom of nothing to do, and impaired vision and, above everything else, the imposition on his family. To him it was wrong in essence. It was an injustice, and worst of all, he could do nothing about it.

He had been living with Peggy at Katanning since his fall at Narrogin, but in 1963 she and Walter Grant moved to the 'Wandgee' property they had bought in 1959 at Green Range, some 60 miles from Albany. After this move, although during the last three years of his life he returned to Wandgee for a couple of months of every winter, Mitchell spent most of his time at Nedlands with Joan, who ordered her life very much to fit in with his requirements: to bed with the sun and, as always, up betimes: playing with his great-grandchildren and delighting in the visits of old friends, although sometimes he did not remember who they were. Apart from his younger sister Nora (Mrs Rae), he was now the sole surviving member of his family.

Nora was known as 'Auntie Sis' from the affectionate name of 'little sister' which Mitchell had bestowed on her in her childhood. She was his favourite: she was a constant visitor to him in Nedlands, and spent hours at a time with him. She was one whom he never forgot.

For some unexplained reason he suddenly gave up smoking: possibly because he no longer had the strength to suck his old pipe. Nevertheless, till general debility necessitated his being placed in a nursing home toward the end of 1966, he still enjoyed his nips of Corio. He became completely bedridden in January 1967 and was admitted to the Home of Peace, Subiaco, but lying in a bed was not the sort of life for Mitchell. He passed away peacefully three weeks later, on the evening of Sunday 19 February 1967.

He was buried in the family grave at Narrogin on the afternoon of Tuesday 21 February, in the presence of the family and representatives of the numerous offshoots. He had outlived his contemporaries, but there were present many of his younger friends of Narrogin days, although now even some of those showed signs of advancing age.

The funeral was preceded by a service in the Church of the Resurrection, on its main altar the magnificent brass Cross presented by the Mitchell family in memory of their mother and inscribed:

A.M.D.G.
in memory of
Ida Marian Mitchell
1874-1954
A Gentle Lady

The officiating priest, the Venerable Archdeacon Frederick Hart, spoke simply and with sincerity. He said in conclusion: 'I have only had the fortune to know this family by repute, but I can say with certainty that forty-eight hours ago the deceased joined the company of his beloved wife, to their mutual joy and comfort.'

At that moment there must have sounded, however silently, for the last time, the symphony of those long and crowded years:

the jangle of harness and the thunder of hooves, the whine of Boers' bullets, the rasp of surveyors' chains and the crack of a polo mallet, the laughter of friends, the voice of a wife and the prattle of children, grandchildren and great-grandchildren. The little boy who had so faithfully delivered his father's horse to 'The Grange' at Bridgetown so long ago had grown up to deliver, just as faithfully, a lifetime of service to his home state.

And now he had reached the end of the road.

1. The branch continued in the old Agricultural Bank office till March 1949 when it transferred to the modernized court house on the corner of Fortune and Federal Streets which had been purchased from the state government.
2. He was allocated a property in the Katanning district, and took up occupation in February 1948.
3. In Furnival Street, now occupied by the Boy Scouts' Hall.
4. Now No. 10 Furnival Street, occupied by Mrs O. Thomas.
5. Bricky had purchased it, on Mitchell's retirement, for £75.
6. Mort Kuhlman went on to play 117 games with Claremont's League side. Dinny Barron played 166 games with Subiaco and then thirty-two with Central Districts, South Australia. He represented Western Australia three times in Interstate matches. Wally Stevens played 108 games with Subiaco. Bert and Wally Walliss were the backbone of Narrogin Imperials for the next eighteen years. Michael Wishart represented Western Australia twice in Rugby Union. Ross Grant made the 'Great Southern Twenty': the local equivalent of an all-Australian team.
7. Janet (Topsy) Wishart married David P. Fischer on 23 January 1959, and their first child, Jane, was born in February 1960.

20
Epilogue

Well, Mitch, now it's finished; have you been correctly described? Did it need thousands of words to cover the subject, or could it not have been stated simply: 'You were a man's man in anybody's language, a gentleman by birth and nature'?

Maybe you were sensitive as to your standard of education. Whilst education is important, it is useless without intelligence, and even intelligence does not rate highly unless combined with wisdom, and in these latter qualities you were singularly blessed.

A rugged individualist, yes, a type the times bred and which our nation needed so much, the type which we are now finding less and less. These days we are developing toward an impersonal society. Initiative is moulded by the welfare state, and personality by the influence of age or income groups. Individualism of the kind which I have tried to portray in you will eventually become no more than a matter of history—a dictionary definition.

The justice you administered, rough though it appears now, was typical of your ilk—tempered with rare understanding and spiced with a touch of humour.

The question may pose itself—did Mitchell's reported diffidence sometimes smack of swimming with the stream? Never. There may have been times when you paused for a while, sniffing the air, as it were—seeking direction. Then you struck out again as manfully and doggedly as ever. Your pay cheque was not a substitute for the desire to be helpful to fellowmen.

A 'rough diamond' you certainly were, but with many facets, each reflecting its own strong brilliant light. How many of the facets of your life and nature have been recorded, how many gone unnoticed? Could not all have been summed up in one remark from your old staff?

'Harry', I enquired, 'I'm writing a biography of Charlie Mitchell. You used to work with him; what do you know of him?'

'Nothing but good, and if you print one word to the contrary, I'll personally sue you.'

If I am accused of over-emphasizing your role, or giving your various undertakings undue importance, then let whatever is in excess be given to those of your era to whom I owe so much, and to whom I have had no other means of saying thanks.

There are two regrets: one is that you could not have been questioned at depth. How much more we would have learnt; but what a furore there'd have been once you had suspected the motive! How we'd have been sent about our business!

The other is that you did not live to read what people thought of you. Yet if you had, I know what your reaction would have been. *Hmph—doesn't know what he's talking about. Never ever did.*

I trust, Mitch, that this would be one time you were not right.

If you can fill the unforgiving minute
With sixty seconds' worth of distance run;
Yours is the Earth and everything that's in it,
And—which is more—you'll be a Man,
My son!

Index